CHARLES DUPIN
(1784-1873)

AND HIS

INFLUENCE
ON
FRANCE

CHARLES DUPIN
(1784-1873)

AND HIS

INFLUENCE
ON
FRANCE

The Contributions of a Mathematician, Educator, Engineer, and Statesman

MARGARET BRADLEY

CAMBRIA PRESS

Amherst, New York

Cambria Press
20 Northpointe Parkway, Suite 188
Amherst, NY 14228

Cover image is the author's portrait of Charles Dupin, based on L. L. Boilly's lithograph (1820) in the Paris Academy of Sciences.

Library of Congress Cataloging-in-Publication Data

Bradley, Margaret, 1933-
Charles Dupin and his influence on France, 1784-1873 : the contributions of a mathematician, educator, engineer, and statesman / Margaret Bradley.
p. cm.
Includes bibliographical references and index.
ISBN 978-1-60497-751-6 (alk. paper)
1. Dupin, Charles, baron, 1784-1873. 2. Dupin, Charles, baron, 1784-1873—Influence. 3. Intellectuals—France—Biography. 4. Statemen—France—Biography. 5. France—Intellectual life—19th century. I. Title.

DC255.D86B73 2011
944.06092—dc22 [B]

2011000416

To the memory of the late François Perrin, and to Ivor Grattan-Guinness, an ever-present help and guide

Table of Contents

Illustration

Preface

The Ecole Polytechnique was set up in Paris in 1794 as a response to an emergency. The various purges that had followed from the French Revolution of 1789 left many of the civil and military engineering organizations short of competent civil engineers and military officers. The purpose of the school was to train a new generation to fill the gap, and as quickly as possible; the students were taught the basics of many techniques (hence the school's name), especially in mathematics, in a three-year course (reduced to two years in 1799). Then the graduates would continue with more specialist studies in one of several civil or military engineering schools.

Leading scientists, especially mathematicians, were recruited to the Ecole Polytechnique as professors, graduation examiners, and lecturers, but the main key to its success was its policy of selecting the most talented students from around the country. The scale of success was surprising; until the late 1810s, the school produced a remarkable cohort of dozens of graduates who took prominent places in the sciences and technological fields, several achieving international repute.

Many of them held chairs or other posts in institutions of higher education or administration. Some went back to the Ecole Polytechnique. For example, as a founding professor of mechanics and analysis, Gaspard Riche de Prony had Augustin Louis Cauchy as a student in the mid-1800s, and a decade later, graduation examiner de Prony worked with Cauchy as a professor there.

One of the most charismatic of the graduates of the Ecole Polytechnique is the subject of this book. Charles Dupin (1784–1873) was a talented mathematician with a special interest in geometry. He followed Gaspard Monge (1746–1818), an important figure in the founding of the Ecole Polytechnique. When Monge was disgraced at the Restoration in 1815 as a close follower of Napoleon and died three years later, Dupin quickly wrote his biography.

Graduating from the school in 1803, Dupin did not proceed to a specialist school but pursued his interest in marine engineering and then technology in general. He took initiatives not only in France but also in Greece and especially in Great Britain, where he conducted a remarkably comprehensive study of the country's engineering facilities in the late 1810s that was published in six quarto volumes. But, his main professional commitment was to serve as a professor of geometry and mechanics in the Conservatoire des arts et métiers from 1819 until his death. In particular, until around 1840 in his opening lecture of the course he would discuss some current theme in French education or science in a public address that aroused much interest and often gained considerable reprints.

Dupin took the public stage still further by his election to the Chamber of Deputies in 1827 and his appointment to the Senate in 1852; he even served for a brief time in 1834 as the minister of the marine. He lived a life that was exceptional not only in its length but also in its involvement in public and even political affairs as well as in scientific research. Interest in him has increased in recent years to the extent that a special meeting on his work was held in Paris in 2007. In this first substantial biography of Dupin, Margaret Bradley, already a biographer of de Prony (Bradley 1998), has captured the wide variety of Dupin's activities, including a comprehensive bibliography of his many writings. May some of his contemporary graduates of the Ecole Polytechnique receive such detailed attention, thereby increasing our understanding of this remarkably fruitful and important period in the history of French science and technology.

—Ivor Grattan-Guinness

ACKNOWLEDGMENTS

My thanks are due to the staff of all the institutions I visited during the course of this work. Liliane Gautret was ever ready to help on technical matters, and Paul Tizard was a constant support.

Fernand Perrin and I had planned to produce this book together, but sadly, he died before the first sentence was written. He entrusted to me all of the documents he had collected, and they were invaluable, as is his memory. Ivor Grattan-Guinness was, as always, an adviser and critic second to none.

ABBREVIATIONS

AAS	Archives of the Académie des sciences
ACM	Archives centrales de la Marine, Château de Vincennes
ADM	Archives départementales de la Martinique
ADN	Archives départementales de la Nièvre
AMV	Archives municipales de Varzy
AN	Archives nationales
AP	Archives parlementaires
ASHM	Archives du Service Historique de la Marine, Vincennes
ASMP	Académie des sciences morales et politiques
BN	Bibliothèque nationale
CNAM	Conservatoire nationale des arts et métiers
DPF	*Dictionnaire des parlementaires français*
DRB	Document Romain Baron (Romain Baron [1898–1985], a professor of literature and historian of the Nivernais, produced studies of the personalities of the region. He had a large collection of Dupin family documents. There is a road in Nevers that bears his name).

ENPC	Archives of the Ecole Nationale des Ponts et Chaussées
JEP	*Journal de l'Ecole polytechnique*
MU	*Moniteur Universel*
PJ	Pièces justificatives
PV	Procès verbaux de l'Académie des sciences, Paris
SMM	Service d'études du Musée de la Marine à Paris

Charles Dupin
(1784-1873)

and his

Influence on France

INTRODUCTION

DUPIN, A MAN OF MANY TALENTS AND CONTRADICTIONS

Charles Dupin has been neglected by historians for many years. In 2007, a colloquium took place in Paris to study this interesting man, and the papers that were presented there have been invaluable sources of reference in the preparation of this work. However, no complete biography of Dupin has ever been published. Perhaps it is the very diversity of his work and achievements—so apparent from the wealth of his writings—that have made him a difficult subject for specialists.

A brilliant student at the Ecole polytechnique and a devoted disciple of the celebrated mathematician Gaspard Monge, Dupin began a successful career as a naval engineer. When his duties took him to Corfu in his early years, he championed the use of the Greek language, and above all, he was involved in the foundation of the Ionian academy. He had considerable admiration for Britain and set his sights on visiting that great rival of France to study all that he could and observe and learn whatever might be advantageous to the progress and industry of

France. After meeting Andrew Ure in Scotland, he became increasingly dedicated to the cause of workers' education.

He was a man of courage, and following the 1814 Restoration in France, when so many great men were threatened, he was ready to put his head on the block for his hero Monge and others. However, he was also constantly concerned with his own advancement. The seeds of ambition were sown early in him, perhaps because of his father's contemptuous attitude towards him and a resulting determination to prove himself. He left no stone unturned to use his contacts and further his own multifaceted career.

Leaving behind what could have been a fine future in mathematics and engineering, he became a politician, playing a major role in French education, not least at the Paris Conservatoire des arts et métiers. He was also active on behalf of women and children in industry and was notable in the establishment of savings banks for the workers. He encouraged women to attend his public lectures.

Dupin could not be put in any political category; indeed, his stances often appeared to be contradictory. The champion of workers' education was eventually accused of ignorance of the true needs of the working class. For Dupin, the salient issues were the progress of industry, economic growth, and the wealth of France; the education of the workers was to be directed to those ends. He was accused of paternalism, and his stance on the slave trade in 1848 would serve to strengthen that point of view. It is difficult to reconcile the courageous scholar of his early years with the politician he subsequently became. He achieved his ambitions and reached the heights he sought, being awarded the title of baron and then pair de France.

Dupin was a cultured man with an impressive talent for languages. He was charming, and many opportunities arose for him, but his life was dominated by his determination to achieve honours. In view of his deep admiration for the French poet Jean Racine, it would be tempting to find in Dupin a Racinian *revirement tragique*—a tragic turning point from a mathematician and engineer to a life in politics and self-aggrandizement.

However, as will be seen, his was, in fact, a Shakespearian tragic flaw: 'Vaulting ambition which o'erleaps itself / And falls on the other'.[1]

To portray Dupin, an enigmatic and charismatic character, I have relied very much on his official and private correspondence, reproduced verbatim in the *pièces justificatives* (quotations from and reproductions of original documents), which illustrates his progress from a great scientist, mathematician, and educator to an ambitious politician. His contribution to the history of French social, scientific, and educational development is undeniable. Nothing can diminish his importance as a great man of his time and a subject worthy of study.

ENDNOTE

1. *Macbeth*, act 1, scene 7.

Chapter 1

A Brief Introductory Survey of Dupin's Life

This is a brief sketch of Dupin's life and achievements, all of which will be expanded upon and detailed in the ensuing chapters.

Pierre-Charles-François Dupin was born on 7 October 1784 in Varzy, Burgundy, France. His parents, local celebrities, already had a one-year-old son, André-Marie, who was destined for a varied life in the law and politics. A third son, Philippe, to be born in 1795, would become a successful lawyer. Their father, Charles-André Dupin, seigneur du fief de Coeurs according to the baptismal certificate of his second son, was to carry out duties of a somewhat controversial nature during the revolutionary years. He became a member of the Assemblée législative and of the Conseil des Anciens and then, after a lapse during the Empire period, sous-préfet. Catherine-Agnès Dupin, Dupin's mother, was of a firm character, highly intelligent, and cultured.

Charles, with his lively mind, sought learning very early, and it was his mother who provided his first lessons. The opportunities for education in Varzy were somewhat limited. The small town had had a college for the offspring of the bourgeoisie, but this establishment had been closed, and education throughout France was in a pretty poor state. After his mother's lessons, young Dupin was entrusted to a cleric, the priest of Menou, Abbé Bougon, and a few months later, he was sent to Orléans for the classes of Abbé Genty, word of whose reputation had reached the Dupin family. This man was not very perceptive, and his first contact with his young pupil was unfortunate: he gave Dupin to understand that he was too young to study geometry. However, Dupin was to prove to his future teacher that he was well able to follow the courses at the school, and he proceeded to demonstrate a rather difficult theorem.

Overcoming all obstacles, Dupin entered the Ecole polytechnique in 1801, placed second on the list of successful candidates. Among the rest were two other notable savants, the physicist Pierre-Louis Dulong (1785 to 1838) and the geometer Olry Terquem, the future editor of the *Nouvelles annales de mathématiques* (New annals of mathematics). Graduating from the Ecole polytechnique in 1803, Dupin proceeded to the Ecole du génie maritime[1] after participating with some of his Polytechnique schoolmates in the construction in front of the Invalides of a landing craft called *Polytechnique*—destined, as they hoped, for the invasion of England.

After his time at the new school (1803–1805), the young engineer was sent to various ports, Genoa and Antwerp among them, but his first really important mission was to Corfu, which was restored to France with the other Ionian Islands by the Treaty of Tilsit (1807). Dupin arrived on the island on 23 February 1808, having narrowly escaped death at sea.

At Corfu, his official function was to be a naval construction engineer, but his true character began to emerge. In addition to being a mathematician and physicist, the educator in him revealed itself; he was to remain an educator until the end of his days. In the realm of engineering, he began to study the flexibility of wood and the strength of materials, but he also participated in the foundation of the Ionian Academy. Interestingly, very little mention is made in histories of the role of the French in establishing education in Corfu, the launching of the Ionian Academy being commonly attributed to the British Lord Guilford in 1824.[2] Dupin himself was apparently something of an enigma to the inhabitants of the island; it seemed to them impossible that one man could give lessons in mechanics and physics, hold his post as a naval engineer, direct an academy, and at the same time, demonstrate his prowess on the dance floor.

In 1811, Dupin fell ill and was repatriated and sent to Toulon. There, he made his mark by founding a naval library and participating in the creation of a naval museum. This was to inspire the subsequent establishment, in 1827, of the Paris naval museum, the Musée Dauphin, the forerunner of the Musée de la Marine[3] in Paris. Dupin continued with his work in naval construction, but he was much attracted by Great Britain; his future travels to those islands were to have a profound influence on him. During his visits, he observed the economic conditions then prevailing in England and Scotland, feeling that they could provide a lesson for France.

In 1818, Dupin was received into the Institut de France (Académie des sciences),[4] of which he had been a corresponding member since 1813. In support of his candidature, he wrote, 'I make so bold as to submit myself for the possible vacancy in the *mécanique* section'. (His predecessor had been J. C. Périer [1742–1814], an industrialist and inventor. A brilliant engineer, Périer was mainly responsible for the introduction of Watt's steam engines into France.) Dupin continued,

> The minister for the Navy has appointed me to Paris on active service. If, subsequently, and for reasons impossible to predict, I were offered duties incompatible with my usual residence in Paris, whatever the advantages of those duties, I should renounce them without hesitation, rather than giving up the honour of a seat in the Académie des sciences. (AAS; PJ 1)

Dupin was very soon elected into the Institut, as the Académie des sciences had become. The ballot gave him an absolute majority on 30 September 1818.

The ensuing years would see more and more honours for Dupin. In 1824, he was promoted to officier de la Légion d'honneur and made a baron, at the same time as the chemist Louis-Jacques Thénard (1777–1857). In 1832, he was elected to membership of the Academy of Moral and Political Sciences (in the fourth section of the academy, 'Political Economy and Statistics'), recently reestablished by the academic politician François Guizot (1787–1874). In 1834, Dupin was appointed a government minister, although for a period of only nine days, 10 to 18 November 1834 (ACM).

The year 1819 marked a new period in Dupin's life—that of an educator and economist. Appointed to a chair at the Conservatoire des arts et métiers on 25 November 1819, he began in 1820 (his course was inaugurated on 2 December) to teach, mécanique appliquée aux arts. Then, in November 1824, his new course in *géométrie et méchanique* [*sic*] *appliquées aux arts* (geometry and mechanics applied to the arts and crafts) began, with more content than the previous course. It must be emphasized that in this establishment, Dupin was not simply a professor; although he would have been content just teaching geometry and mechanics, he was also an organizer. Moreover, his ambition was to produce equivalent courses in the French provinces, and it was for this reason that he addressed to the minister for the navy a long communiqué entitled *Enseignement de la classe ouvrière* (Education of the working class). Dupin aimed to create education for the workers in fifty French towns. He began his treatise thus:

> If industry amongst our neighbours remained static, we could abandon ours to the slow pace of its natural progress, and there would be no worries in seeing routine halt the march of this progress. But, to the West, the North and East of France, rich and densely populated nations are ceaselessly seeking to increase the means of improving their useful arts. (Originally published in the Annales de l'industrie manufacturière, reproduced in Dupin 1826b; PJ 2)

To his role of educator, one must also add his political role; in Dupin, the two were closely linked. In 1827, shortly before the elections of April 1828, when he was député for the Tarn, he produced a treatise entitled *Situation progressive des forces de la France depuis 1814* (The progressive situation in the strength of France since 1814), followed by an account, *Forces électorales à la fin de 1827* (The voting population of France at the end of 1827). In the latter, he wrote,

> France has the greatest efforts to make to raise herself, through elementary education, to the level of peoples whom we regard as ignorant. In this respect, we are beneath the Irish and the Austrians; forty years ago, seven million French people could read; twelve million can today, but 26 million should be able to do so. Let us work to provide for 14 million of our fellow citizens the first and most essential knowledge. In ten years, we must do more than our forbears did in 40. (Dupin 1828b, 1; PJ 3)

Dupin was much attached to family life. He often enjoyed the company of his two brothers, either at Chitry-les-Mines, a small canal port in the Nièvre, or at the Château du Réconfort, a property he acquired in 1825. The Château du Réconfort was not far from Corbigny, also in the Nièvre, where he spent his last years. He married rather late in life, at the age of forty-six, a young woman of twenty-three, Rosalie-Anne Joubert, the only daughter of tradespeople in Angers (Perrin 1983, 18). There were three children of their union: Catherine-Rosalie-Anne-Marie Dupin (1832–1904), who married Comte Pierre du Hamel de Breuil in 1855; Charlotte Dupin (1834–1907), who married the Marquis Gaston-Félix de Lentilhac in 1858; and Charles-André Dupin (1835–1869). In his will, Dupin stated, 'It is my deeply felt desire that one of my grandsons should bear my name and my title of Baron, to keep alive the memory of all that our family has done for this dear country' (AN MC/ET/XLVI/1170, 4 December 1871). The son died childless, and the Dupin name has now died out.

It has proved to be very difficult to find much information about Dupin's family life, but one letter from his wife, addressed to Mme. de Prony and dated 22 June, indicates that the couple shared friendships. The year is not given in the letter, although its content indicates that it was written around 1850. It refers to Dupin's involvement in the

Chamber of Deputies and with an exhibition, probably the British Great Exhibition of 1851, for which Dupin was the president of the French committee (see chapter 14.5). The letter is one of sympathy for the loss of Mme. de Prony's sister-in-law:

> Your sister-in-law is yet another victim of the dreadful scourge afflicting us at the moment … As soon as M Dupin has free time, he will visit you, but apart from the chamber, the exhibition committee of which he is president increases his workload … (BN Naf, 3282)

The 'dreadful scourge' was probably cholera, of which there was a major outbreak in France in 1832 and a second in 1849. Dupin was a close friend of the great engineer Gaspard de Prony, to whom there will be many references in subsequent chapters.

In his little Nivernais[5] region, Dupin involved himself in local duties, doubling his national ones; he was the mayor of Saizy, a commune in the Nièvre, and conseiller général[6] for the Nièvre town of Tannay. Faithful to his native town of Varzy, he gifted to it the family library of more than two thousand volumes.

Shortly before his death, Dupin wrote to Louis Pasteur, who was thirty-eight years his junior, 'Allow me to send you my warmest congratulations for your admirable research on alcohol fermentation … the way looks clear to create from this work a new era in science to which you owe your well-earned reputation' (ASHM CC7 779; PJ 4). This letter is dated 28 August 1872.

A few months later, the Academy of Sciences received a letter dated 18 January 1873, signed by the Count du Hamel de Breuil, Dupin's son-in-law, announcing Dupin's death: 'My father-in-law died this morning …, the result of the rupture of a probe in his bladder … [he] died fully conscious, patient, resigned, comforted by his faith and confident in God's mercy' (AAS; PJ 5).

Dupin was buried in Clamecy, a small town in the Nièvre, and speeches were made at his graveside—one by General Arthur J. Morin (1795–1880), representing the Conservatoire des arts et métiers, and the other by Charles Lévêque, a professor at the Collège de France, on behalf of the Académie des sciences morales et politiques. The mathematician and Ecole polytechnique professor Joseph Bertrand paid homage to Dupin in an *Eloge historique* (In praise of Charles Dupin), which he read to the Academy of Sciences on 2 April 1873.

Thanks to the family of Hamel de Breuil, Dupin's memory was kept alive (Perrin 1983, 19). This family carried out restoration work at Saizy, and one of Dupin's grandsons created in 1920 the Charles Dupin prize at the Academy of Sciences, to reward the author of a book or paper on geometry, mechanics, or maritime navigation.[7]

I shall now proceed to examine in detail all of these varied aspects of Dupin's life and contributions.

ENDNOTES

1. For the organization of the Ecole polytechnique and its relationship with the application schools, see Bradley 1976c and Fourcy 1828.
2. Frederic North, Earl of Guilford (1766–1827). On a visit to Corfu in 2009, this general belief was confirmed. The seat of the Ionian Academy is there to be seen, but mention was made only of Guilford and the British in creating the first Greek university. Similarly, with regard to the fortifications, the pamphlets do not refer to the French contribution.
3. The Musée de la Marine was installed and directed by Gustave Zédé, a naval engineer, to whom some have attributed the invention of the first submarine (in 1888).
4. The academy had been abolished in 1795, when the Institut de France was established.
5. The Nivernais is a traditional province of France, like Poitou, Anjou, or Bourgogne.
6. Roughly the equivalent of a British county councillor.
7. This was originally a septennial prize, but it has been awarded somewhat irregularly.

CHAPTER 2

DUPIN'S BACKGROUND AND FAMILY

2.1. THE EARLY YEARS

Although Dupin left the Nivernais when he went to the Ecole polytechnique, his origins and family history provide information about his formative years. His birthplace, Varzy, lies between the Val de Loire and the granite lands of the Morvan. It was in the neighbouring town of Clamecy that his father resided in the course of his professional duties. Dupin's memory seems to have been eclipsed by that of his elder brother, Dupin *aîné* (eldest son), both at regional and national levels, probably because scientists have long been neglected by historians who have little or no knowledge of the sciences. In contrast, mathematicians have studied Charles Dupin in depth and have developed his work, but that is of interest to a limited number of readers. The Dupin of this study is little known. His childhood and adolescence coincided with a troubled period in French history. His family was disrupted by national events, but it seems that the character of his parents and brothers enabled them to adapt to circumstances. I shall now look at those other members of his family.

2.2. DUPIN'S FATHER

Charles-André Dupin (1758–1843) was descended from an ancient, bourgeois Varzy family which had produced many clerics, businessmen, and lawyers (Perrin 1983, 35–43). Born in Clamecy, he became a lawyer in 1779, but deafness prevented him from practising. He moved to Varzy in 1781, where he assumed various official posts. On 7 September 1791, he was elected by the Nièvre department to the Legislative Assembly. Returning to Clamecy, he was suspected of being a moderate when he came up against the extremist Jacobins who had taken over the district. In January 1793, he was listed as suspect—'tarnished by conduct unworthy of a citizen' (Baron 1978)—and was put under house arrest for several months. Freed on 17 August by Joseph Fouché,[1] who was on a visit to Clamecy, Dupin considered it prudent to retire to Varzy, where his wife resided. This did not, however, prevent him from being incarcerated two more times, first at Nevers in October 1793 and then for two months at Pressures (the Château de Pressures, between Clamecy and Varzy, which was used as a prison during the Revolution; both the Abbé Bougon and Dupin père were imprisoned there). Dupin père was set free in January 1794 by J. A. Lefiot, Fouché's successor. Returning to Varzy, he realized that if he were to be free of worries, it would be wise for him to keep a comparatively low profile in politics. He joined the minority of Jacobins of the Société populaire, which in the main was of a moderate tendency, and he became their official spokesman until the fall of Maximilien Robespierre in July 1794 (see Baron 1978).

This period of relative calm enabled Charles-André Dupin to devote his attention to the education of his sons, André-Marie, who became the famous Dupin *aîné*, and Pierre-Charles-François, the future Baron Dupin, the subject of this work; in 1794, they were aged eleven and ten, respectively. However, this period did not last long, and C-A. Dupin was again persecuted, not as a moderate but because he had supported the Clamecy terrorists. According to Baron (1978), this is an obscure time in his life. It is not known whether he spent it at Clamecy or whether he was appointed a judge in Varzy at the district tribunal a month before 9 *thermidor* (27 July 1794)[2] only to lose his post during the *thermidor* reaction.[3] It may be that he stayed at home with his wife. In any case, on 3 *floréal an* 111 (22 April 1795), a district representative arrived in Varzy to proceed with the disarmament of the Jacobin extremists, Dupin among them, 'for having thrown himself in with the terrorists, and obstinately preached their principles after coming out of Pressures' (Baron 1978). The municipal authorities declared that Dupin had not been seen in Varzy since 9 *thermidor*.

During the Directoire period, things settled down for C. A. Dupin, who assumed new posts. He was appointed government commissaire at the Clamecy tribunal, and the following year, government commissaire in the central administration of the Nièvre. In year VIII (1800), the Nièvre department elected him by 95 of 115 votes as député du Conseil des cinq-cents,[4] an assembly which was dissolved shortly afterwards as a result of Bonaparte's coup d'état. C.-A. Dupin became a member of the Corps législative in 1800, and then he was appointed a delegate to the General Inspection of the Gendarmerie. In 1814, after Waterloo, he was the public prosecutor and then sous-préfet of Clamecy (2 August 1815). He made his mark in that post, according to Thuillier, who stated, 'The préfets received poor support from the sous-préfets between 1815 and 1830. Apart from Dupin *père*, who was very competent, … the conseillers de préfecture and the préfets seem to have been nondescript administrators' (1976, 93). This was Baron Dupin's father.

2.3. The Eldest Son, André-Marie-Jean-Jacques Dupin, 'Dupin *aîné*' (1783–1865)

The eldest son of Charles-André Dupin remains indisputably the best known of the three Dupin sons. It was his statue that was raised in their hometown of Varzy, and it is he who has aroused the most public interest and the attention of authors. His professional life in the law and politics was certainly dynamic, and his career may be more easily understood than the scientific activities of his brother Charles. Commonly known as 'Dupin *aîné*', Charles-André was a prestigious lawyer and wrote some hundred books on the law. As a magistrate, he was a prosecutor at the court of appeal for almost thirty years. Elected député at the end of the Restoration period, he had been a member of the private council of the Duc d'Orléans, the future King Louis-Philippe, since 1820; the king's accession to the throne was greatly assisted by Dupin *aîné*. In 1827, he was reelected député for the Nièvre, and so he remained without interruption until 1851. He presided over the Assemblée from 1832 to 1840 and from 1849 to 1851. Rallying belatedly to the Empire, he resumed his duties as a prosecutor in 1857 and continued in this role until his death. He was a member of the Académie française, the Académie des sciences morales et politiques, and grand cordon of the Légion d'honneur. During his reign, Dupin *aîné* was Louis-Philippe's close adviser (see Perrin 1983, 44–55).

In his *Mémoires* (4 vols., 1855–1863), Dupin *aîné* spoke of the important role his father played in his education. He did not often attend the colleges of Clamecy and Varzy (which were, in any case, closed for several years during the Revolution), but his father concentrated on his eldest son's education to the point of exhausting him. There is no doubt that this serious, basic education stood him in good stead throughout his long career. As for his career, he was a first clerk with a Paris lawyer, a doctor of laws at the age of twenty-three, and then he set himself up as a lawyer in the Latin Quarter of Paris. However, he was not unanimously admired. In 1848, Charles Robin wrote,

> M. Dupin is shortsighted with pretensions to far-sightedness and he does not know what is going on around him. An inveterate compiler, he has collected here and there, witticisms, repartee and quotations which he uses all the time and out of context ... Brusque of manner, sharp, trivial and low in his

> language, M. Dupin has the appearance of some kind of Danube peasant, but scratch the surface and you find the smooth skin of a courtisan. Excessively vain, he has had the effrontery to claim himself to be a Demosthenes in court, Cicero at the bar and Cato the elder in the field. As a lawyer, he has pleaded for and against with marvellous facility; as a magistrate, he is uncompromising on privilege; as a politician, he drifts with any current. (2:406–407)

An anonymous rhyme was written in a similar vein:

> Every power in turn can say, he's one of us,
> Stale bread [*du pain*, or 'Dupin'] for those out of favour, fresh rolls [*du pain mollet*] for the rest,
> Presented with a chair of office, he accepts,
> And all Paris says: it's stale bread [*du pain rassis*, a play on the words *Dupin rassis*, 'Dupin reseated']. (ADN 4 J, packet 1 of 2; PJ 6)

The fact remains that Dupin *aîné* was a highly respected lawyer and is still considered to have left his mark on his profession. In the preface to his *Du droit d'aînesse* (1826), dedicating the work to 'mes frères' (my brothers), he spoke of the influence of his parents on him and his brothers and how they inculcated in them the 'desire to be useful to our fellow-citizens and our country'. He wrote the book because of an 1826 bill that reestablished the rights of the eldest child. Dupin was strongly opposed to this attempt to return from a civil state to an ecclesiastical one—the Ancien Régime.

There was great affection between the future Baron Dupin and his elder brother, and Charles greatly respected Dupin *aîné*, whom he addressed as 'Mon cher Dupin'—the big brother who often protected him from their father's wrath. In 1814, young Charles wrote to his brother, 'It seems that our father has particular and special reasons for not writing to me. No news for a year. *Non sum dignus* [I am unworthy; this expression precedes the Latin Roman Catholic Communion]'. His brother replied that their father was very busy, to which Charles retorted, 'I don't take bladders for lanterns'[5] (DRB; PJ 7). It will be seen that the father was often irritated by Charles. In fact, he called him 'mathematicomaniac', and in a letter to his eldest son in 1813, he complained, 'When we speak of family matters, this maniac replies mathematics, *Institut* and Monge' (DRB).

Charles wrote affectionately to his father, who rarely replied; if he did, his tone was quite the opposite of affectionate, as will be seen in future chapters. However, as his father was approaching the end of his life, he seemed to wish to be reconciled with his second son:

> If you considered my great age and the disaster which has come to worsen and hasten the course of my life,[6] we could use at least the little time that is left to me to make our contact more frequent and, for that, so much more comforting. Sooner or later, you will experience in your turn all that this doubling of filial affection gives to a father. (DRB; PJ 8)

2.4. The Influence on Charles Dupin of His Father and Mother

In 1782, Charles-André Dupin had married a third cousin, Catherine-Agnès Dupin, who was five years his junior. She brought a dowry of her house in Varzy and the fief de Coeurs, an ancient property. She was the daughter of Jean-Jacques Dupin, a lawyer, and Anne-Vincente Robin. According to Baron (1978, 64), the marriage was hardly a success. Agnès Dupin had her charms and had studied the finest literature, which was unusual for a woman in her day. She was of strong character, with an independent and original mind. To her eldest son, she communicated a ready wit, which became one of his special qualities. She was hugely ambitious for her sons and took charge of their teaching when her husband was in prison. Her pride was to be in motherhood of the *trois Dupin* (the three Dupin boys).

However, Madame Dupin was a very dominant woman and distinguished herself during the Revolution by the violence of her denunciations. Her husband found that patience was the only answer to her. There was discord in the household from the start, and according to a document in the Varzy archives,

> the father of the three Dupins, born in Clamecy, took up his usual residence there, but because of domestic strife, which even ended at one point in a divorce suit [22 *pluviose an* V (10 January 1797)], the mother lived separately from her husband [the divorce was finalized just before 1810]. Because of this separation, it was in Varzy that she gave birth to the three boys. (AMV; PJ 9)

This explains why they were not born in Clamecy, where their father lived.

The tension between the spouses probably resulted from the father's Jansenist—and therefore religious—persuasion, whereas the mother was a disciple of the eighteenth-century philosophers and the principles of enlightenment. On 17 *fructidor an* XII (30 August 1804), the father wrote to his eldest son to tell him that the mother had visited the town with the youngest son, Philippe, then aged nine. She had taken him everywhere except to visit his father, who had learned that she wanted to send Philippe to Orléans at a cost of one thousand francs a year. He wrote, 'It is utterly stupid. It is as mad as it is wicked. She wants to be absurd, even at her own expense. I have done everything possible to have nothing with which to reproach myself'. He chose to add a spiteful comment about his second son, Charles: The great engineer is no better' (ADN; PJ 10).

In spite of their own disharmony, the Dupins had a deep and happy influence on their two eldest sons. With just a year between them, the boys enjoyed the same childhood. The difficult climate engendered by the Revolution, their father's sufferings, and the trouble between the parents could have had an ill effect on the two boys, but both the father and the mother were fine teachers. During their father's imprisonment, their mother took up their instruction and

> taught them ancient history giving them to read the *Abbé* Barthélémy's *Les voyages du jeune Anacharsis* and Robin's Roman History ... When Dupin *père* finally returned to his family,

> he again took up the reins to educate his children. He had enjoyed an excellent education at the Paris *Collège Sainte-Barbe*, and he applied the methods practised in that highly reputed establishment. He taught them the humanities, of which the Latin authors were the essential ingredient, and the study of the great French writers of the century of Louis XIV. (Baron 1978)

Until the end of 1799, Charles Dupin had received only his parents' teaching, supplemented by lessons with the Abbé Bougon, the priest of Menou, near Varzy, who had lent his voice in support of the Revolution. The abbé was condemned to deportation in 1798 but saved by Dupin père, who offered him asylum. He was pardoned in 1799 and joined the Dupin household as a tutor.

After the 18 *brumaire* coup d'état (9 November 1799), Dupin père, a member of the Corps législatif, took his eldest son to Paris, with this same abbé as his mentor. Charles, then aged fifteen, left for Orléans to prepare for the Ecole polytechnique entry examination (for the details of this examination, see Bradley 1974). There was then no *baccalauréat*, and there were no 'special' classes. Charles was to attend the Ecole centrale, one of the departmental schools inspired by d'Estutt de Tracy (1754–1836).[7] Their creation was proposed by the député Joseph Lakanal in 1795, and their emphasis was to be on science. In his new school, Charles would be taught by Abbé Louis Genty du Haume (1763–1817), a member of several learned societies. Charles would recall his time there in a speech he gave to the Ionian Academy:

> An adolescent, was educated in a public school, far from his family, whom he would not see again. However, the end of the school year brings the award of prizes. Already the magistrates and all that the town can offer in dignitaries are assembled, in the presence of a huge crowd of people. The laurels are distributed. The young student had one, and thought only of his little literary glory, drunk with his first experience of honour and distinction; he receives his prize; he turns round: what does he see? His mother, his good mother, who had made the journey in secret to take pleasure in her son's happiness … Prizes, laurels, spectators, ah! I forgot them all; pride gave way to nature, I saw

> only my mother, my gentle mother, and I flew into her arms. (ADN 4 J)

Still very young, Dupin was thus completing the first chapter of his life; he had received a somewhat Spartan education but had greatly benefited from his intellectual family background. He would reach the heights with his sharp intelligence, eclectic but methodical mind, and enormous capacity for work. Like his older brother, he was not without his critics, mainly for his support for the Second Empire. However, at the time of the Restoration, he took a great risk to defend Lazare Carnot and his eldest son, Sadi Carnot (1796–1832). Lazare Carnot was popularly known as the 'organizer of victory'. In 1793, when France was suffering invasion and was threatened on all sides, he organized mass mobilization to save his country. The phrase was coined by the politician Jean-Denis Lanjuinais and soon taken up by the people. Sadi Carnot, a staunch republican, had joined Napoleon during the Hundred Days campaign (for more on Sadi Carnot, see Fox 1978). Lazare Carnot and Jean-Denis Lanjuinais, a lawyer who had fought for the rights of the *tiers état* (the third estate, or the common people), would, together with Dupin, attack the reactionary ministry of Villèle in 1827 (see chapter 12.5).

Dupin also publicly defended his guide and mentor Gaspard Monge against the attacks of the new Restoration administration in 1815. His devotion to Monge and his desire to be the great geometer's 'spiritual son' were described by Belhoste, a professor of the history of science at the University of Paris (Christen and Vatin 2009, 87). Lazare Carnot and Monge were the only two members of the first class of the Institut to be excluded from the first class of the Institut, and Dupin's support for them became public. Lazare Carnot insisted that Dupin should not publish the defence he had made of him to avoid serious repercussions.

Endnotes

1. Joseph Fouché (1759–1820), a Girondin. He was an organizer of the movement for dechristianization and opposition to the *noblesse*.
2. The Revolution had produced a new calendar, with the names of months reflecting the climatic conditions and the months divided into three *décades,* periods of ten days.
3. The thermidor reaction of 27 and 28 July 1794 resulted in the fall of the Jacobins, victory for the Centre, and an end to the Terror.
4. The Conseil des cinq-cents, the legislative assembly established by the constitution of year III (1795).
5. A colloquial expression meaning 'I'm no fool'.
6. He was losing his sight.
7. One of d'Estutt's aims was to create a scientific language in the interests of providing clarity in scientific research.

CHAPTER 3

DUPIN'S WORK AND CONTRIBUTIONS IN HIS EARLY YEARS

3.1. DUPIN'S PUBLICATIONS

The diversity of Dupin's writings bears witness to the breadth and depth of his interests (see primary bibliography). Indeed, it is difficult 'to determine where the "real" Dupin is, beyond the kaleidoscope of his many activities' (Fox 2009, 7).[1] Apart from all of his scientific and mathematical achievements, he had a profound knowledge of literature and a level of proficiency in Ancient Greek (see chapter 5). Such eclecticism is quite a rarity in the modern days of specialization, but this was not the case in the past. Dupin's friend, the great engineer Gaspard de Prony (see Bradley 1998), for instance, wrote music and poetry with which he and his very musical wife provided entertainment for their guests.

Nevertheless, Dupin's father, always so very sceptical about his second son, wrote to his number one son on 15 July 1818,

> I have started to read Charles's writings. He is good when he is on his own territory. That is where his reputation lies, rather than in his excursions into literature. Having so long despised all that is not numbers and lines, he writes his mother tongue like a foreigner who has just learned it. (DRB; PJ 11)

There is probably some truth in the observations of Dupin's father, although he was generally prejudiced against Charles. Charles had concentrated his attention on the sciences, and there were gaps in his knowledge of grammar. He had been long absorbed by his engineering work, and in his many papers, technical considerations took precedence over style. It is on his work as a mathematician, scientist, naval engineer, and educator that I shall concentrate in the following sections, although more will be said of his literary interests.

Charles's scientific career began when his studies at the Orléans *école centrale* (the Orléans central school) equipped him for entry to the Ecole polytechnique at the age of just seventeen (*an* X, 1801). His year of entry was to provide other savants, the physicist and chemist Pierre-Louis. Dulong (1765–1838) and the geometer O. Terquem among them; Antoine Marie Augoyat, Dupin's great friend, was also a first-class civil engineer.

What was the Ecole polytechnique at this time? There are many misconceptions about it, the worst being that it was a creation of Napoleon Bonaparte.

3.2. The Origins and Early Years of the Ecole Polytechnique

The original idea for the creation of a school to provide a systematic education for all types of engineers may be attributed to J. E. Lamblardie (1747–1797), the director of the Ecole des ponts et chaussées. The chemist Antoine François Fourcroy (1755–1809) produced the report that resulted in the creation of the Ecole centrale des travaux publics, as it was originally called. However, it was the distinguished geometer Gaspard Monge who is considered to be the father of the school, because he produced the programme of studies in collaboration with the mathematician Jean-Louis. L. Lagrange (1736–1813), the mathematician Jean Baptiste Joseph. Fourier (1768–1830), the chemist Claude Louis. Berthollet (1748–1822), and the celebrated engineer Gaspard-Clair-François-Marie Riche de Prony (1755–1839).

The new institution was to be free and open to any young men who showed potential, but in its first ten years, only 19 percent of the students came from families of *sans-culottes* (the 'trouserless ones', as the very poorest members of society were known by the Revolution). (For details of the early history of the school, see Bradley 1974, 1975, and three 1976 papers). Previously, mathematicians and engineers had received their education at, for example, the Ecole du génie in Mézières, the Ecole militaire in Paris, the Paris Ecole des ponts et chaussées (founded in 1747), the Ecole d'artillerie at Châlons sur Marne, or at the Ecole des élèves ingenieurs constructeurs de vaisseaux (founded in 1765), the forbear of the Génie maritime. The idea of the new school was to unite all of these schools under one roof, which would make such an establishment more easily controllable; it was placed under the auspices of the minister of the interior.

The school began life as the Ecole centrale des travaux publics, and it was not until 1795 that it took the name Ecole polytechnique. In 1805, it was installed in the former Collège de Navarre, where it remained until 1976, when it was transferred to Palaiseau. Although it was in the hands of the minister of the interior, he was obliged to consult the minister of war and education on important issues. In 1799, a *conseil de perfectionnement*, or council for improvement, was created with the director and members of the teaching staff to establish the programme of studies and to ensure coordination between the Polytechnique and the reestablished specialist engineering schools (it had soon become apparent that these were necessary because it was not possible to create different specialist engineers in one three-year course). Dupin's year of entry was during the interim period before the school was militarized by Bonaparte (by the imperial decree of 27 *messidor an* XII [16 July 1804]), who referred to the school as his 'hen of the golden eggs' (this is a literal translation, the English term being 'goose'). Dupin sat for the entrance examination in Paris, one of many centres that was established throughout France for the purpose. His Ecole polytechnique record describes him as having

> hair and eyebrows light brown, low forehead, medium nose, brown eyes, small mouth, short, round chin, oval face, height 1m. 64. Military service, admitted to the school for the service and construction of ships. Parents' domicile: his father, a member of the *Corps législatif*, at Clamecy (Nièvre). (EP archives, *dossier* Dupin, fiche de matricule, the entry record)

According to the register of the Conseil de perfectionnement of 15 November 1801, Dupin came first among the students who took the exam in Paris but second overall, although, according to Perrin (1983, 87), the few brief biographies would have it otherwise, with Dupin coming in first. On 31 May 1803, Dupin was granted leave to work on the building of cannon boats. He was admitted to the Génie maritime on 3 October 1804.

The programme of studies at the Ecole polytechnique included no literature and not even French-language studies until François-Guillaume J. S. Andrieux became the professor of grammar and literature in 1804.[2] What was life like at the Ecole polytechnique in Dupin's time? The hours of study were long, but the students enjoyed considerable independence and freedom when they were not in class. They were provided with a uniform and a modest grant, with which they were expected to pay for their board and lodging. Very much a product of the Revolution, the school was probably the first in the world to help all of its students financially. From its inception, 'respectable' citizens, known as *pères sensibles*, were chosen to take students into their homes. They were to provide a simply and appropriately furnished room and food at their table and were enjoined to take care of their charges as good parents would with their own children. Interestingly, however, homes were not eligible if they had more than one staircase.

From the time of his entry to the Polytechnique, Dupin revealed a remarkable aptitude for mathematics, and he involved himself in work leading to his *cyclides*, which will be studied in chapter 3.4. Some scholars believe that this work marked him for distinction at this early age; according to an article by Maurice d'Ocagne in the *Figaro* of 23 November 1933, the results of his studies 'were published even whilst their author was still a student at the *Polytechnique*'.[3] Dupin's prowess is the more surprising because the mathematical skills that were required for the entrance examination to the school were of a somewhat modest nature, according to the *Rapport sur la situation à l'Ecole polytechnique, présenté au Ministre de l'Intérieur par le Conseil de perfectionnement, établi en exécution de la loi du 25 frimaire an VIII* (Report on the Situation at the Ecole polytechnique Presented to the Minister of the Interior by the Council for Improvement, Established According to the Law of 16 December 1788) (AP; 16 December 1799). However, according to this report, the examination consisted almost entirely of mathematics:

1. Arithmetic.
2. Algebra, including the solving of equations of the first two degrees, the general composition of equations, the demonstration of Newton's binomial theorem but only in the case of complete positive exponents; methods for finding the commensurable factors which may be contained in an equation, to solve numerical equations by approximation, to eliminate the unknowns in equations of all degrees [*sic*].
3. Elementary geometry, including rectilinear trigonometry and the use of logarithms for solving triangles.
4. The main properties of conic sections.
5. Static mechanics, mainly applied to the equilibrium of simple machines.

The report specified that 'the candidates must write, under the dictation of the examiner, some French sentences to show that they know how to write their language correctly'.

Some years after his graduation from the Polytechnique, Charles Dupin was to express his criticism of the gulf between the entrance examination and the teaching at the school in a letter to his friend Mathieu (Claude-Louis? An explanation follows):

> I have suggested to Hachette [J.-N.-P. Hachette, a professor of mathematics at the Ecole polytechnique] that he produce for the *Ecole polytechnique*, a book on the following plan ... Between the knowledge of applied analysis of the candidates for the *Ecole polytechnique* and Monge's lectures, there is an enormous gulf, of which Hachette's lectures bridge only the smallest part: Hachette deals only cursorily with second degree surfaces. He considers only the tangent planes and makes abstraction of their curves,[4] and his lectures should sow the seed for all the great theories of Monge's lectures; coming to him [Monge], and before applying the calculus of partial differentials, they [the students] should have all that simple differential calculus teaches on surfaces, and Hachette says nothing of all that. Also, you know how few students are capable of following Monge's lectures (ADN; PJ 12)

It is noteworthy that Dupin's interest in this letter was, characteristically, in teaching.

His great friend was probably the astronomer Claude-Louis Mathieu (1783–1875), who was a student at the Ecole polytechnique at much the same time as Dupin. It is apparent that Dupin and Mathieu were very close. Mathieu was born in Mâcon in 1783, the son of a carpenter. He joined the Polytechnique in 1803, the year of Dupin's graduation, and then proceeded to the Corps des Ponts et Chaussées. He moved to the Bureau des Longitudes in 1809, became an astronomer at the Paris Observatory in 1809, was elected to the Academy of Sciences in 1817, was appointed a temporary professor at the Collège de France, and was a Polytechnique examiner until 1863. He worked on meridian measurement and collaborated with François Arago, who became his brother-in-law in 1821.

3.3. Mathematics in France at the End of the Eighteenth Century and the Start of the Nineteenth Century

When Charles Dupin joined the Ecole polytechnique in 1801, French mathematics had many famous men working in the subject, for example, Gaspard Monge, Joseph-Louis Lagrange, Pierre-Simon de Laplace, Charles Fourier, Adrien-Marie Legendre, Lazare Carnot, Jean Nocolas Pierre Hachette, Siméon-Denis Poisson, André-Marie Ampère, and Carl-Friedrich Gauss[5] (see Perrin 1983, 99–117, and Grattan-Guinness 1990, where there are many references and very full details of the work of all of these mathematicians). Augustin-Louis Cauchy, one of the creators of modern analysis, came later; his *Cours d'analys* (Analysis course) appeared in 1821.

Gaspard Monge (1746–1818) was the author of a work on descriptive geometry which was first published in 1799. As a result, he was appointed *répétiteur* (teaching assistant) to the Abbé Bossut at the Ecole de Mézières, where he had been a student. In 1768, at the age of twenty-two, he replaced the Abbé Bossut. In 1780, he was elected to the Académie des sciences, and during the Revolution, he took an active part in the defence of his country. He was minister for the navy during the revolutionary years and participated in Bonaparte's Egyptian expedition (see Taton 1951).

Monge was the organizer of the teaching plan at the Ecole polytechnique, where he was a professor; he also taught at the Ecole normale.[6] In addition to descriptive geometry, he produced numerous works on infinitesimal geometry and the theory of curves. He laid the basis for the differential geometry of surfaces and developed, in a new and original way, analytical geometry (which was called at that time 'algebra applied to geometry'). He also studied the problem of the solution of equations of partial derivatives of the first order and in 1771, surfaces according to their mode of generation. Dupin, one of Monge's favoured pupils, owed him a great deal. Bonaparte expressed his appreciation of Monge: 'Citizen Monge', he declared, 'one of the members of the *Commission des sciences et des arts*, is famous both for his knowledge and his patriotism. He won respect for the French through his conduct in Italy' (report by Bonaparte, on the treaty of Campo-Formio, to the Directoire exécutif, 27 vendémiaire an 6 [18 October 1797]; AP). More will therefore be said about Monge in later chapters.

Le Comte Louis de Lagrange (1736–1813) was the author of extremely important works on variations and developments in series and above all, the *Mécanique analytique* (Analytical mechanics) of 1788. He established a system of general equations on the movement of a material system which are well known to students of rational mechanics. Lagrange was a professor at the Ecole normale as well as the Turin Military School. He established in particular a theorem on the movement of solid bodies, with which Poisson's name is also associated (the Lagrange and Poisson movement).

The Marquis Pierre Simon de Laplace (1749–1827), a professor at the Ecole normale from its inception in 1794, was the author of a cosmological theory that is contained in his *Exposition du système du monde* (Explanation of the system of the world; 1796), but his most notable works were the *Mecanique céleste* (Celestial mechanics) and memoirs on the calculus of probabilities. He also engaged in research on calometric measurement, the adiabatic transformation of gases, capillarity, and electromagnetism (see Gillispie 1997).

Laplace's work was made known in Britain by, among others, J. Playfair (1748–1819), J. Brinkley (1763–1835) in 1803 (see Smith 1976), and Mary Somerville, who published her first paper, 'The Magnetic Properties of the Violet Rays of the Polar Spectrum', in the proceedings of the Royal Society of London in 1826. She went on to be elected to the Royal Astronomical Society in 1835. Widely travelled herself, she entertained Dupin during his second study visit to Britain (see Patterson 1983).

Baron Joseph Fourier (1768–1830) made his name in the sphere of trigonometric series and an analytical theory of heat. He taught analysis at the Ecole polytechnique, where he was one of the school's first teachers, although he was no longer there when Dupin joined it. Fourier participated in Bonaparte's Egyptian expedition with Monge and Berthollet.

Adrien Marie Legendre (in fact, 'Le Gendre' is correct) (1752–1833) was the author of works on plane and spherical trigonometry, a theory of elliptical functions, and a theory of numbers. He exposed the method of lesser squares before Gauss did. Legendre was a professor at the Ecole militaire, a graduation examiner at the Ecole polytechnique, and a member of the Bureau des Longitudes. His *Elements of Geometry* remains a classic.

Lazare Carnot (1753–1823), a military engineer, was a député in the Legislative Assembly. Although he was an ardent revolutionary, he dissociated himself from Robespierre. He was outspoken in his hostility to Napoleon I, but when France was in danger of invasion in 1814, he was active in the defence of Antwerp. He was famous for his organization of the army in southern France in 1793, contributing to his description as an 'organizer of victory'. A professor at the Ecole polytechnique, he was the author of several works on pure mathematics and mechanics, for example, *Essai sur les machines en général* (1783, Essay on machines in general), *Géométrie de position* (1803, Geometry of position), and *Essai sur la théorie des transversales* (1806, Essay on the theory of transversals).

The Restoration period was a bad time for Dupin's guide and mentor, Monge. During the first Restoration, the government decided to purge the Institut, and thus it was that Monge, Carnot, and Guyton-Morveau were expelled from the first class (the Académie des sciences). Then, after the Hundred Days (18 June 1815), Monge, when he was at the Institut on 3 July 1815, was threatened by the government and had to go into hiding at the home of François Arago, the renowned mathematician, physicist, and politician of Catalan origin. Finally, the Institut de France was completely dissolved, and in particular, Monge and Lazare Carnot were struck off from the *mécanique* (mechanics) section and replaced. On 13 April 1816, by royal decree, the Ecole polytechnique was temporarily disbanded.[7]

All of these developments spelled disaster for Monge, who died not long afterwards on 28 July 1818. The government went so far as to refuse the Polytechnique students the right to attend his funeral, but on the following day, they went en masse to his grave at the Père-Lachaise cemetery. Monge had been a well-respected tutor who had had direct contact with the students, often meeting them after class. His style was very much that of a British college tutor, and he left an indelible mark on the Ecole polytechnique. In the repressive climate of the Restoration, Barnabé Brisson (1777–1828), who was one of Monge's first students at the Ecole polytechnique, spoke in his praise. He had become a *ponts et chaussées* (bridges and highways) engineer and had married Monge's niece. Louis Guyon, a Polytechnique student, also spoke at Monge's funeral, as did Charles Dupin, who was then a Génie maritime engineer. They did not hesitate to run the serious risk of publishing, each one, an *éloge* (eulogy) of their former professor (Brisson 1818; Guyon 1818; Dupin 1819c). Moreover, Dupin recalled Monge's greatness on 2 September 1849, when he presented *Éloge de Gaspard Monge* (In praise of Gaspard Monge) to the Académie des sciences (Dupin 1849b).

Monge remains remembered for his work on descriptive geometry, analytical geometry, and second-degree surfaces (see Taton 1951 for a comprehensive account of Monge's life and work). One of his quirks was his lack of interest in publishing either his lectures or his mathematical discoveries. In his 'Discours prononcé aux funérailles de Hachette' (Funeral oration for Hachette [18 January 1834]), Charles Dupin said of Monge,

> The characteristic of the great geometer, whose pupil and assistant Hachette felt honoured to be, was his deep repugnance to publishing the demonstration of a great number of remarkable properties and ingenious methods revealed in his lectures, either at the *Ecole du génie de Mézières*, or at the *Ecole polytechnique* … M. Hachette rendered an invaluable service to science by undertaking the work of editing and publishing several works under his name or that of M. Monge.

In similar spirit, in his *Essai historique sur les services et les travaux scientifiques de G. Monge* (Historical essay on the service and scientific work of G. Monge), Dupin commented,

> Monge had the shortcoming of being reluctant to commit to paper the fruit of his meditations. It could be said that the great truths of science were the object of his research merely for the pleasure of rising to them and contemplating them in their grandeur. (1819c)

It may be added that Monge often lost the papers that were lent to him; Dupin was to complain about this, as well as about Hachette for the same reason, as will be seen.

There is no doubt that Monge had a profound influence on Dupin, not only through the quality of his teaching but also through his kindness. Monge appreciated the lively intelligence of his pupil and their intellectual affinity. However, it should be added that the warmth between them, and also that between Dupin and Lazare Carnot, owed a great deal to the friendship of the two savants with Dupin's mother, dating from the early days of the Revolution. Similarly, Dupin's father had established in Paris a friendship with Carnot, who greatly encouraged Dupin in his studies. Subsequently, during politically difficult times, Charles was able to show his gratitude. Lazare Carnot had to seek exile during the Restoration (under the order of 24 July 1815); Dupin undertook to defend him and asked to support his cause before the Chambres should they be called to pass judgment on him. Moreover, Dupin produced a defence entitled *Du jugement de M. le lieutenant-général Carnot* (On the judgment of Lieutenant-General Carnot; 1815), but as was seen in chapter 1, Carnot opposed its publication because of the danger that would be involved for Dupin.

3.4. Dupin's Early Mathematical Work and Discovery of the *Cyclides*

Dupin's great interest was in mathematics and pure geometry, as much because of the influence and prestige of his professor Monge as by his natural inclination. To be precise, he was continually concerned with the notion of curvature and in particular, the lines of curvature, a subject of Monge's work. Thus it was that he evolved the notion of the *cyclides* (see Perrin 1983, 118–147). This discovery, among his many works, was not the one that resulted in the most major applications, but it was chronologically his first, and therefore it attracts attention because of the young age at which Dupin developed the idea. Unfortunately, he left no dated notes on the early development of the *cyclides*. However, it is possible to establish an approximate date for his work by looking at his 1822 'Les applications de géométrie et de méchanique [*sic*]' (Applications of geometry and mechanics) and volumes 1 and 2 of Hachette's *Correspondance sur l'Ecole impériale polytechnique*[22] (Correspondence on the Ecole impériale polytechnique; a regular publication for the use of the students, created by Hachette in 1804) (Hachette 1804–1816). The *Journal de l'Ecole polytechnique* makes no mention of the *cyclides*, a family of surfaces characterized by the fact that their lines of greatest curvature and their lines of least curvature are exclusively circles.

In his 'Applications de géométrie et de méchanique [*sic*]', Dupin, after a paragraph on optics entitled 'Properties of *cyclide* surfaces and second degree surface curves', stated,

> We shall prove, firstly, that there is a family of surfaces whose characteristic property is to have only circles for lines of greatest and least curvature: this is why we shall call these surfaces *cyclides* … The analysis of the results of this *Mémoire* is in the *Correspondance polytechnique* … the *Mémoire* itself has not been published. (*4ème mémoire*, 200)

In other words, only the analysis of the paper had been published in the *Correspondance*.

In volume 2 of Hachette's *Correspondance*, there is an article by Dupin entitled 'Mémoire sur la sphère tangente à trois ou quatre autres' (On the sphere tangent to three or four others), dated from Pisa, 1 October 1812, in which he stated, 'The publication of my *Mémoire* is still delayed, so I withdrew it to improve it and then, during a hurried trip to Belgium, I lost it' (Hachette 1804–1816, 1813, 420–425). He mentioned an article in the first volume (April 1804–March 1808), adding, 'I knew that M. Hachette referred to my work on the questions dealt with in this *mémoire*, analyzing one of the questions it contained. This analysis is in the second section of the *Correspondance sur l'Ecole polytechnique* (volume 1)'. Indeed, the first volume contains an article by Hachette entitled 'Travaux de l'Ecole, mémoire sur le contact des spheres' (Hachette 1804, 17–18).

In his article in the second volume, Dupin stated,

> Ten years ago, I took up a subject so often raised and never, it seems to me, as generally as it deserves. I submitted my solutions to Monge and I had the good fortune to receive from such a geometer, great encouragement.

This reference to ten years earlier—that is, 1802—recalls the time when Dupin was a student at the Ecole polytechnique and had discovered a family of surfaces known as the *cyclides*.

Dupin's *cyclides* have been the subject of various analytical rather than geometrical developments—in particular, by Théodore Moutard (1827–1901), a professor at the Polytechnique and the Ecole des mines; in 1863 by the German E. Kummer (1810–1893), a professor at the Universities of Breslau and Berlin; and in 1873 by Gaston Garboux (1842–1917), who provided a generalization, expanding the notion of the *cyclide* to every fourth-order surface, including the Maxwell circle of infinity. Clerk Maxwell of Cambridge referred to Dupin's work in 1846 (1846), as did Chandru, Dutta, and Hoffmann in the *Quarterly Journal of Pure and Applied Mathematics* (1989, 111–126).

After graduating from the Ecole polytechnique, Dupin's scientific activity and his work on the *cyclides* covers a comparatively short period of his life, lasting until around 1816, when he was aged thirty-two. The sources for Dupin's story so far are the *Journal de l'Ecole polytechnique* and his two articles, 'Essai sur la description des lignes et des surfaces de second degré' (On second degree lines and surfaces, 1808, vol. 7, *cahier* 14, 45; 1807a) and the 'Expériences sur la flexibilité, la force et l'élasticité des bois' (Experiments on the flexibility, strength and elasticity of wood, 1815, vol. 10, *cahier* 17, 137; 1813a), which will be explored further in the sections studying Dupin's work in naval construction and particularly in chapter 4. More of his earlier work may be found in Hachette's *Correspondance sur l'Ecole polytechnique*, as follows:

> Vol. 1, *Notice sur les travaux de l'Ecole* (Notice on the work of the school), years 11 and 12, 8 and 9;
>
> 'Travaux de l'Ecole', *Mémoire sur le contact des sphères* (On the contact of spheres) 17–28; and
>
> 'Géométrie descriptive', *Analyse d'un mémoire sur les surfaces du second degré*, (Analysis of a paper on second-degree surfaces) (144–148).

After 1816, and his return from Britain, Dupin was about to launch out in different directions, as will be seen in subsequent chapters. However, I shall first look at the interim period and the consequences of Dupin's early work.

3.5. The Impact of Dupin's Early Mathematics

It has been seen that Dupin published in the *Correspondance sur l'Ecole polytechnique* his 'Essai sur la description des lignes et des surfaces du second degree '. This was very much a repetition of his 'Mémoire sur les surfaces du second degree', written in *nivôse an* XII (a period from 22 December 1804 to 20 January 1805). He had also published his 'Expériences sur la flexibilité, la force et l'élasticité des bois' in the same journal in 1811.

What was the *Correspondance*? Hachette was a professor at the Polytechnique and also taught at the Paris Faculty of Sciences. In addition, beginning in 1810, he added teaching at the Ecole normale to his duties. In an interesting preface to his first volume (volume 1, 1804), he explained the destination of the *Correspondance*:

> The greatest pleasure we may enjoy is, without doubt, that of meeting everywhere where public works are to be organized, former students of the *Ecole polytechnique*, fulfilling with zeal

> and success their engineering duties, loving the school and its professors, having one for the others that lasting love founded on the generous sentiments which inspire a large family devoted to the arts and sciences ... All feel the need to maintain a correspondence with the alma mater and this need is even stronger when they are far away: to respond to a wish so generally expressed, I propose the publication of correspondence, aimed mainly at the students of the *Ecole polytechnique*; this correspondence would make known the particular work of the professors, the developments in the teaching methods.

On his 'Mémoire sur les surfaces du second degré' (1807a), Dupin commented in the *Correspondance*,

> M. Dupin was charged with a very difficult military mission, aside from all his other duties, when he wrote this paper; he managed to correct it subsequently, but only on a four hundred league journey.[8] Therefore, one should not hope to find in this paper the order of things to arouse interest, nor the correctness and clarity which is justifiably expected from this sort of work: geometers will judge whether the ideas in themselves are worthy of indulgence or oblivion (Antwerp, *ventôse* year 13, end of February and beginning of March 1805 (1804–1805)

Towards the end of the paper, Dupin pointed out that

> the method of description, the subject of this paper, gives, for tracing the lines of curvature of second degree surfaces, a very simple means which has potential for application to the crafts, and particularly to stone-cutting, which, until now, has had no geometric means of describing by continuous movement the lines of curvature which are, as is known, the backbone ... in ellipsoidal arches.

Again, Dupin's preoccupation with application, as well as with curves, emerges. The paper was presented to Monge, who was similarly concerned with practical applications. It should also be mentioned that the paper was drafted by Dupin when he was working as an engineer in shipbuilding, although he seemed by nature to be more of a geometer than an engineer. In military terms, Dupin was then an officer of the Génie maritime, but only sous-ingénieur de 2e classe, not yet a fully qualified engineer. He drew much from the work of the Swiss mathematician Leonhard Euler (1707–1783) as well as, of course, Monge.

In the same volume, of the Correspondance (volume 1, 218–225), there are comments on Dupin's 'Sur la théorie des déblais et des remblais' (On the theory of cuttings and embankments) to the effect that Dupin's work was an extension of Monge's researches on the same subject. Monge had observed that the most advantageous system of transport in route construction is that in which 'the sum of the products of the mass of elements transported by the space covered by each is minimal'. Dupin's paper studies the case where roads cross in a cutting or an embankment. He reminded the reader of the importance of considering pure geometry, which, he acknowledged, may seem to be far removed from the projected aim—that is, the construction of roads in the most economical manner possible. He explained:

> Let us imagine the case where they [the roads] are obliged to change course abruptly; suppose, for example, that every return journey of the workers or the transport carriages, must pass through a given point. Such would be the case for distant journeys where only one journey a day could be made: all the workers, horses etc. would have to return by each route covered, to their homes or workshops ... It would obviously be ridiculous to try to assign to each element, or every minute load, the route it should cover. But by dividing the routes according to the workshops, as is always done in major projects, in determining the furthest routes that separate the transport from the various workshops, it will suffice that the transporting be done following intermediary directions to the most distant limits, these being necessarily and adequately indicated.

In volume 2 (January 1809 to March 1813, 86–87), one finds 'Sur la courbure des surfaces' (On the curvature of surfaces), wit the editor's explanation, '(two theorems to be demonstrated, submitted by Charles Dupin'. Hachette commented, 'One of his [Dupin's] friends wrote to me that, whilst awaiting publication, this could be included in the *Correspondance*' (1813, 86–87). This was a work on the curve and osculation of surfaces. In the same volume (387–396), one finds Dupin's 'Analyse de plusieurs mémoires de géométrie' (Analysis of several papers on geometry), read to the first class of the Institute on 14 December 1812. This analysis mentions, in particular, the *indicatrix* (indicator) and the theorem of conjugated tangents (391). Dupin explained that it was towards the end of 1805, during a journey between Holland and Italy, that he began this research. He continued it in 1806 when he was in Genoa and in 1807 when he was in Toulon, 'still during the brief moments of leisure time that my duties allowed me'. In 1808, he was to leave for the Ionian Islands, and he explained,

> I found myself obliged, through perhaps unique circumstances, to devote myself to my mathematical research, almost without help, advice or even books … Endlessly occupied by a thousand different matters and constrained my state duties, it is the work of an engineer that I herewith present and not the fruit of the meditations of a savant. (Hachette 1813, 86–87)

Although Dupin enjoyed pure mathematics, he had a lively interest in practical and useful applications (the stability of ships, cuttings, and embankments). As regards the indicatrix, many scholars consider Dupin's discovery in this sphere to be more important than the *cyclides*.

In volume 3 of Hachette's *Correspondance* (January 1814 to January 1816, 132), one finds Hachette's 'Analyse appliquée à la géométrie' (Analysis applied to geometry), in which he introduced extracts from one of Dupin's papers and one of another *polytechnicien* (polytechnician), Michel-Ange Lancret. Dupin's paper is on the subject of conjugated tangents (renamed by Hachette 'reciprocal tangents'). Lancret's paper is on double curves, curves of constant precession (Lancret 1806).

Amédée Mannheim (1831–1906), a professor at the Ecole polytechnique, commented, 'Euler … led Dupin to the conic he called the *indicatrix* … The theorem of conjugated tangents, extremely useful, has numerous applications' (1894, 332). Mannheim's work also contains an undated theorem of Charles Dupin's: 'When a right is displaced so that three or four of its points remain on three given planes, a fourth point of this right describes an ellipsoid whose centre is the meeting point of the given planes' (175).

3.6. *The Développements de géométrie*

The *Développements de géométrie … (pour faire suite à la Géométrie descriptive et la Géométrie analytique de Monge)* (Developments in geometry [to follow the descriptive and analytical geometry of Monge]) was Dupin's first published book, appearing in 1813 and first announced at the end of an article in the *Correspondance sur l'Ecole polytechnique* entitled 'Analyse de plusieurs mémoires de géométrie' (Analysis of several geometry papers). This was presented to the first class of the Institut on 14 December 1812. Two weeks later, after a favourable report by the three commissaires, Carnot, Monge, and Siméon Denis Poisson, it was proposed to insert it in the *Recueil des Sçavans* [*sic*] *étrangers* (Collection of work of external scholars).

On 1 February 1813, Dupin presented two other papers, approved in their turn by the first class and deemed worthy for the same publication. Three months later, the *Développements de géométrie* appeared; it was effectively a collection of these papers and was dedicated to Monge. It should be observed that the work was essentially theoretical, and the applications would wait nine years to appear in Dupin's *Applications de géométrie et de méchanique à la Marine, aux Ponts-et-Chaussées etc. … pour faire suite aux Développements de géométrie* (Applications of geometry and mechanics in the Navy, Bridges and Highways etc. … to follow the Developments in geometry) (see chapter 13.2).

In 1813, Charles Dupin was only twenty-nine years old. It was in that year that he was elected correspondent of the first class of the Institut and awarded the title of chevalier de la Légion d'honneur. As for his military grade as captain in the Génie maritime, that corresponded to sous-ingénieur de 1ère classe, and he would wait until 1819 for the title of engineer (third class).

In his dedication of the work to Monge, Dupin explained,

> During these long years I have had to spend in almost barbarous regions, far from home … what I most regretted was the friendship of those illustrious men, whose kindly precepts raised me above myself and whose presence was for my courage an inspiration which, every day, renewed my easily exhausted strength. (1)

The 'almost barbarous regions' were Corfu and the Ionian Islands, of which more will be said in chapter 5. Dupin couched his dedication in somewhat pompous terms, but to his friends, he used a different tone. In a letter from Genoa to Mathieu on 10 May 1806, for example, he was much more down to earth: 'I should like to know if that old rascal Monge will be good enough to reply to me on my work "Des déblais et des remblais"; I'd swear that he won't' (ADN; PJ 13).

To digress a little, it is of historical interest to look at Dupin's postscript to this letter, in which he asked Mathieu to stamp his letters: 'Just think that for three *sous* letters to soldiers are stamped and that yours costs 18 here. Soldiers are penniless beggars. Stamp yours or send them to my father who will forward them to me without paying postage'.

In his *Essai historique sur les services et les travaux de G. Monge*, which appeared during the year following Monge's death, Dupin wrote,

> Other questions no less interesting for the arts, but which Monge, considering them to be too complicated for simple descriptive geometry, decided against treating too deeply … are those concerned with determining, according to certain laws, surfaces which have with others, contacts of second order. That is what one has tried to achieve in a work following the *Géométrie descriptive* and Monge's *Géométrie analytique* which appeared in 1813, under the auspices of that great geometer. (1819c)

Dupin was referring to his own *Développements de géométrie* in a fine example of self-publicity.

For the Institut, a commission comprising Carnot, Monge, and Poisson reported, on 14 December 1812, on Dupin's first three papers on geometry and analysis. They described the work as 'relating to the curves of surfaces. His main object is to make this theory more elementary and above all more easily usable in the numerous possible applications' (Procès-verbaux de l'Académie des sciences).

In all of his scientific work, Dupin aimed to be as clear as possible, directing it in particular to the students of the Ecole polytechnique and the Corps du génie. He was to become a professor at the Conservatoire des arts et métiers, where his teaching would be inspired by the desire to be understood by students who had no real basic knowledge of science (such as working men and overseers). He was guided by two main considerations: social usefulness and practical application (although, as has been seen, he often worked very much in the realms of theory). His scientific work took place during a relatively short period; Taton (1951) referred to the 'importance of the new results he achieved within the space of ten years'. However, it must also be said that Dupin systematically pursued his own professional advancement and social climbing. In the next chapter, I shall look at his work as a naval construction engineer and his further progress up the professional ladder.

ENDNOTES

1. A collection of papers on Dupin, edited by Christen and Vatin, *Charles Dupin (1784–1873)*, is listed in the bibliography with a full list of contributions.
2. F. G. J. S. Andrieux (1759–1833). His lectures were very popular.
3. This is, in fact, incorrect.
4. Note the importance of curves in Dupin's thinking.
5. Carl Friedrich Gauss (1777–1855), the director of Göttingen Observatory and the author of *Disquisitiones Arithmeticae*, which is still regarded as one of the most influential books in mathematics.
6. The Ecole normale was created in 1794 by the Convention. Joseph Lakanal, speaking on behalf of the Committee for Public Education (Le comité d'instruction publique) in 1794, explained its purpose: 'Citizens from all parts of the Republic, already instructed in useful branches of knowledge would be called so that they might learn from the most skilled teachers, the art of teaching' (AP).
7. The significance of the school's closure may be judged from the publication of the complete royal decree of 13 April in the London *Times* on 19 April 1816.
8. A journey to Italy he took at the end of 1805, probably with the following itinerary: Antwerp, Brussels, Valenciennes, Péronne, Paris, Sens, Auxerre, Avallon, Saulieu, Chalon sur Saône, Mâcon, Lyon, and Turin, taking account of the lines of communication of the time.

Chapter 4

Dupin as a Naval Construction Engineer Until 1816

4.1. The Background

By the treaty of Campo-Formio (17 October 1797), France was permitted to annex the Austrian Low Countries and to take possession of the Ionian Islands. Above all, the country still had the possibility of directing all of its military forces against Britain. After the peace of Lunéville in 1801, Bonaparte envisaged the installation of ports on the Channel coast for embarking troops for England, and he moved the Ecole spéciale du génie maritime to Brest (see Bradley 1994). Vice Admiral Lord Nelson, off the coast of Boulogne aboard *Medusa*, wrote to Admiral Duckworth[1] warning him of the French preparations to invade their country and insisting on his (Nelson's) determination to thwart the French plans.

The signing of the Peace of Amiens (on 25 March 1802) was rapidly followed by hostilities between France and Britain. While the French dockyards were hives of activity rebuilding the fleet and Bonaparte was set on preserving his domination of the Mediterranean, Britain sent France an ultimatum (on 26 April 1803) demanding the evacuation of Holland and Switzerland. When the time expired, an embargo was imposed on French ships (on 16 May 1803). The following day, the Peace of Amiens was broken. The ports of Brest, Rochefort, and Toulon were blocked by British squadrons, and France was cut off from its overseas possessions. It was then that Bonaparte returned to his idea of invading Britain and decided (on 14 June 1803) to concentrate a force of one hundred and fifty thousand soldiers on the Channel coast. The camp of Boulogne was created as the headquarters for this army, spread over various ports. These ports were to build a flotilla of flat-bottomed boats for the transport of troops. At the same time, five other camps were created: Compiègne, Saint-Malo, Bayonne, Ghent, and one in Holland.

In these preparations, the Ecole polytechnique became involved in a role that was more symbolic than effective, but that indicated both the position of that young institution in France and the enthusiasm of its students. On 26 May 1803, just nine days after the rupture of the Peace of Amiens, there appeared in the *Correspondance sur l'Ecole polytechnique*, paragraph 3, page 9:

> 'Inspired by this sacred love of our homeland deep in the heart of every good Frenchman, we feel it our duty to contribute from our grant, until we can pay personally'. During the morning, a sum of 4,000 francs has been collected and deposited the same day, in cash, in the Public Treasury… They asked permission to donate an armed ship of which they themselves would be responsible for the construction. The government acceded to their wish; a cannon-boat received the name *La polytechnique*, *n*° 287.

In the same paragraph, it was reported that the first consul had ordered that thirty students (of whom Charles Dupin was one) of the Ecole polytechnique should be prepared by a month's preliminary instruction to be employed in naval construction. The government order was instituted, and a month later, the thirty chosen students were distributed in the ports, or *départements*. Charles Dupin was destined for the Ecole des constructeurs de vaisseau, the Génie maritime, and he was to make a rapid transition from mathematics to naval construction. However, in the course of his naval engineering duties, he would still pursue his mathematical researches, making discoveries that were perhaps even more important than the *cyclides*.

4.2. The Ecole du Génie Maritime

At the end of 1803, Charles Dupin began his studies at the Ecole du génie maritime, having already received some instruction in that area at the Ecole polytechnique. In 1765, a school of *ingénieurs constructeurs de la marine* (naval construction engineers) had been created, with its seat in Paris at the Louvre. Closed in 1793 because of its attachment to the Ancien Régime, it reopened in 1795 with a new name, l'Ecole des ingénieurs de vaisseaux, and was again situated in Paris, in the rue Saint-Dominique in the Loménie de Brienne building. The school's director was the mathematician and physicist J. C. Borda (1733–1799), a former officer of the Génie militaire who had transferred to the navy (see Mascart 2000).

In 1799, the Corps des ingénieurs de la marine took its final name, becoming the Ecole spéciale du génie maritime (not to be confused with the Ecole navale, which did not yet exist). A ruling on 7 *floréal an* VIII (27 April 1800), completed by another on 7 *thermidor* (26 June), militarized the naval engineers. By the terms of this ruling, an officer of the Génie maritime in each division, under the orders of the division commander, would be responsible for work related to the repair and maintenance of ships. Moreover, a superior officer was to be aboard with each squadron of fifteen vessels.

Imperial conquests led Napoleon to transfer the school to the coast, in keeping with the demands of his naval policy. The Génie maritime moved to Brest from 1801 to 1810, then to Antwerp from 1810 to 1814, falling back to Toulon in 1814 after the military reversals of the Empire. It is therefore probable that Dupin studied in Brest, although no evidence has been found to that effect. A letter from Dupin to his friend

Mathieu, dated 13 *brumaire an* XII (23 December 1803), was sent from Compiègne, where one of the six camps created by the First Consul (on 14 June 1803) was situated:

> This sublime Bailly [J. S. Bailly (1736–1793), an astronomer who was executed during the Revolution] writes with such interest and charm, it is with such nobility and genius that he propounds on the great phenomena of nature and the history of their discovery that one cannot weary of admiring, reading and re-reading him. You will sense all the charm of the style, all the beauty of the ideas and the merit of his works when you compare them with those of others … Laplace's *Système du Monde*, for example, will almost always please you and, sometimes even will fill you with enthusiasm if you read it without knowing Bailly. But if you already knew him, the charm would immediately diminish. You would love Bailly's simpler dignity, less complex and yet more eloquent and majestic. (ADN; PJ 14)

This is perhaps a far cry from naval matters, but it illustrates Dupin's many interests, mathematics, engineering, and literature along them.

Sadly, there is an absence of records of Dupin's progress at the Ecole du génie maritime. After Toulon, the school returned to Paris, where it remained. In 1970, it became a part of the Ecole nationale de techniques avancées (ENSTA).

4.3. France, 1803 to 1805, and Its Navy

To understand Dupin's activities and travels during these years, I shall now look at the extent of the French territory and the considerable importance to France of an invasion of England. In 1790, France had 83 departments but, from 1793 with the annexation of foreign territories, the number had increased to 130 by 1809. In addition to the departments of which France now consists, there were four on the left bank of the Rhine (Sarre, Rhin-et-Moselle, Mont-Tonnerre, and Roër), several in the Alps (in particular Léman), six in Piedmont (Marengo, Pô, Doire, Stura, Sesia, and Tanaro), and nine Belgian départements (Jemmapes, Escaut,

Lys, Ourthe, Dyle, Meuse-Inférieure, Deux-Nethes, Sambre et Meuse, and Forêts).

The plan for the invasion of England has been studied by many historians (see Bradley 1994). Tulard commented,

> Was the idea of a conquest of England by the French army quite realistic? Did Bonaparte believe that such an expedition was possible or was it just a means of intimidation to bring pressure on the British cabinet to negotiate? Was the choice of Boulogne, in the first hypothesis, a good one? And did France have a navy capable of this enterprise? So many questions which have constantly attracted totally contradictory answers. (1970, 272)

It was in 1805, after years of hesitation and preparations, that the emperor seems to have decided on making an invasion attempt. Tulard noted,

> From now on, Napoleon sees that he is condemned to succeeding. His plan rests on two elements, squadrons and flotillas. The flotilla is to ensure the crossing of the Channel from the ports of Boulogne, Calais, Etaples and Ambleteuse … In fact, this flotilla of mediocre quality, seems unable to face stormy seas and make headway amongst the English ships keeping watch on the Channel. Forfait ill, Admiral Bruix disillusioned and Decrès, Minister for the Navy, concerned above all with behaving like a good courtisan, can be blamed for the deficiencies of the flotilla. (1970)

From the technical point of view, this flotilla was an original idea in naval construction at the beginning of the Empire. At Boulogne, there were several engineers who had recently graduated from the Polytechnique with the highest marks (e.g., Charles Dupin, Jean-Baptiste Marestier,[2] Jean Hubert, Pierre Moreau, and Jean Tupinier[3]), and they made an active contribution to the project of the flotilla, which was completed within a year (construction took place from 1804 into 1805). In spite of the inevitable inadequacies of a hastily prepared fleet, on its first sortie it defeated an English squadron 'before the eyes of the army acclaiming this heroic battle', according to the account in the *Ecole polytechnique Livre du centenaire* (vol. 2, 1895).

4.4. Charles Dupin's Activity After Leaving Boulogne Until His Departure for Corfu

There is comparatively very little information recorded about Dupin during the years between his leaving Boulogne and his departure for Corfu. It has been seen that he was at Compiègne on 23 December 1803, beginning his studies at the Génie maritime. According to the notice about his 'Essai sur la description des lignes et des surfaces du second degré', he was at Antwerp in *ventôse an* XIII (the end of February/ early March 1805). He was in Boulogne early in 1805 under the orders of Tupinier, employed in new constructions and the completion of the flotilla, and then, apart from the time he spent in Antwerp, he was in the Dutch ports with military workers. It is also known that after France's Trafalgar defeat (21 October 1805), he made the long journey from Antwerp to Italy; it was during this journey that he began the mathematical researches he described in the *Correspondance sur l'Ecole polytechnique* (vol. 2, 387–396). From a letter to his friend Mathieu of 22 February 1806, it is clear that Dupin was then in Genoa (a town that had recently been annexed to the Empire), where he and several Génie maritime engineers, among them Tupinier, had been sent by Napoleon, who had abandoned the idea of invading England. As an engineer, Dupin worked on enlarging the port and studied the art under the direction of Pierre-Alexandre-Laurent Forfait. As a mathematician, he continued the researches he had begun during his previous journey. Beginning around this period, more material about his activities is available.

Various letters make it possible to establish a picture of Dupin's work in 1806, especially those written to Mathieu, to whom he was obviously very close. In the following section are extracts from five letters Dupin wrote to Mathieu dated from Genoa and Toulon, couched in familiar and witty terms at the same time as they reveal Dupin's tenacity and determination to succeed.

> Genoa, 22 February 1806
>
> This devil of a Monge cannot be found and gives me no reply. Ah well! What can you do, my dear Mathieu, we'll give up and

> have an end to it. I'll have only one regret in all this, and that is to have given you so much trouble to no avail. Since you are at the *Ponts-et-Chaussées*, you must have Monge's paper on cuttings and embankments, *Académie des sciences*, Paris 1781. I have worked a lot on that subject since Genoa. If you would like me to, I would give you my work and you could, if it seemed worthwhile, show it to Prony [director of the Ecole des ponts et chaussées] to whom this work must necessarily be of interest. But how the devil will I ever know that? You're not much of a writer and one can only get news of you, third hand. Admit that that's not your best point. (ADN; PJ 15)

A month later, Dupin was in Toulon and returned to the same questions in another letter to Mathieu:

> Toulon, 17 March 1806
>
> You told me that you had sent the analysis of my paper on cuttings and embankments to M. Delambre [J. B. J. Delambre (1794–1822), the renowned French astronomer and mathematician; he was involved in the measurement of the arc of the Paris meridian between Dunkirk and Barcelona in the interests of the establishment of the metric system]. He has probably lost sight of it and has never spoken to you about it. Regarding this analysis, I was most amazed, two or three days ago, when one of my colleagues, reading the *Moniteur* quoted an article as follows: *Corresp. Polyt.* [*sic*] 'Analyse d'un mémoire de M. Dupin sur les déblais et les remblais etc.'.
>
> I had given this analysis to Hachette to pass it to Monge, when Hachette and Poisson [S. D. Poisson (1781–1840), a French mathematician, geometer, and physicist, known for his work on definite integrals, electromagnetic theory, and probability] came to Genoa. Because Hachette, to whom I had sent, for Monge, a first copy, confessed ingenuously that it was lost and he did not know what he had done with it: it will probably be on the second copy that Hachette will have based his article regarding me. (ADN; PJ 16)

The other letters written by Dupin to Mathieu in 1806, from which I provide extracts, are again dated from Genoa. However, this does not mean that Dupin made no return trips to Toulon during this period, although most of his activity in naval construction was based in Genoa, as there was much to do there. The first extract is informative on Dupin's work in Genoa.

> Genoa, 27 March 1806
>
> You will observe perhaps, my dear Mathieu, that I brought back optics as no more than a particular case of the theory of cuttings and embankments. You must have seen in no. 5 of the *Correspondance*, a fine quality of optics given by M. Malus [E. L. Malus (1775–1812), the 'chief Laplacian optician' (Grattan-Guinness 1990, 473)].
>
> He achieved it by analysis; I demonstrated it by geometry; I took it as far as I could … I have frequently returned to what I had done on the curvature of surfaces. Just think, my friend, that all optics, all stonecutting, the entire theory on cuttings and embankments, etc. make, with the theory of the signs of curvature, a single and vast system of ideas. We should hang all those rascals of invisible and bustling senators who are, however, the best men on earth; I don't need to tell you that I am flattered by the opinion of M. Monge, but I would have thought that the idea of seeking three systems of constantly rectangular surfaces would have been the one he preferred. Here is what I think I have shown:
>
> If three systems of any surfaces are such that they cut three to three at the right angle at each point of their intersection, as would be the case with a group of small cubic boxes, the surfaces of the first system are cut by those of the second following all their lines from one of their curves, and the same surfaces of the first system are cut by those of the third following all the lines of the other curve; thus as soon as one knows a similar group of systems, which will include a given surface, the intersections of this surface and of those of the other systems will give the lines of curvature. (ADN; PJ 17)

In a comment on the work of A. Ribaucour (1845–1893), an engineer and mathematician who studied in depth the 'triple orthogonal systems', the authors stated, 'Finally, we describe all submanifolds carrying a

Dupin principal curvature normal vector field with integrable conullity, a concept that has proven to be crucial in the study of reducibility of Dupin submanifolds' (Dajczer, Florit, and Tojeiro 2007). Another reference to Dupin is made in Corro and Tenenblat: 'As an application of Ribaucour transformations, we characterize the Dupin hypersurfaces which have a principal curvature of constant multiplicity one, as a manifold foliated by (n-1)-dimensional Dupin submanifolds associated by Ribaucour transformations' (2004).

An extract from Dupin's letter to Mathieu of 10 May, referring to 'that old rascal Monge' has been quoted in chapter 3.6 (PJ 13). Similarly, another letter, also of that date and criticizing Hachette's teaching (PJ 12), has been mentioned in chapter 3.2. That extract is followed by these remarks of a pedagogical nature:

> Here is an example of the simplicity that I would endeavour to apply to the most difficult subjects. Suppose I am asked to determine the direction of the lines of curvature of a second degree surface and the radii of curve of the same surface for one of the given points. This is how I do it:
>
> Through the centre of the surface, I draw a plane parallel to the plane tangent to the surface at the given point. I determine the axes of the section of this plane in the surface. The large axis is parallel to the tangent of the main line of curvature at the given point. The lesser axis is parallel to the tangent of the lesser line of curvature at the given point. The square of the main axis of the section, divided by the distance from the centre to the tangent plane at the given point is equal to the greatest radius of curvature. The square of the small axis of the section, divided by the distance from the centre to the plane tangent to the given point, is equal to the smallest radius of curvature. And these results can be obtained by immediate considerations; they bring back, as you can see, the contacts of second order to proportional resolutions, which is much easier. I have dug like a poor wretch on the curvature of surfaces. All that I have found on second degree surfaces I have tried to generalize for any surfaces, this leading me to results which I shall communicate to you in the future. I can immediately and graphically, for example, solve this problem: Any three

> curves cross at the same point on a surface; the curvature of the curves is known; the question is the direction of the lines of surface curvature and the length of the two radii of curvature which belong at this point of the surface. This solution can be applied in stonecutting, which, as you know, is closely linked with the theory of lines of curvature.
>
> When I spoke to you on the *Mémoires de l'Institut* about my work on cuttings and embankments, it was more to have the opinion of M. Delambre than the desire to see my papers in the *savants étrangers* [external scholars], a somewhat mediocre honour. (ADN; PJ 18)

Application was very much in Dupin's mind in this letter, as was facilitating the solution of problems—all very much directed towards the student and practising engineer. However, in the final part of the letter, is there a note of arrogance, or was he really most concerned with the opinion of an influential mathematician who could be useful to him as he furthered his career?

As regards Dupin's return trips to Toulon, there are a few clues. A letter of 31 October 1806 from the préfet général maritime[4] to the minister for the navy and the colonies refers to 'M. Dupin, *sous-ingénieur*, recently arrived in the port of Toulon' (ASHM, CC 7779). A letter from Dupin to his elder brother dated 8 December 1806 also indicates Charles Dupin's presence in Toulon:

> I have arrived after 23 days of travelling and having covered 24 leagues on foot. I was 20 days on board ship, sleeping on the deck without a mattress, but I am very well; that's in the past and I think of it no more. (DRB)

There appears to be no further explanation of this journey. Dupin continued by congratulating his brother: 'I have a thousand compliments to pay you, my dear friend, on your success. Far from surprising me, it fulfils all my expectations and I can only thank fortune for having been so fair to you' (DRB; PJ 19). At the age of twenty-three, Dupin *aîné* had received the title of doctor of law. It was the first doctorate to be awarded by the Paris law faculty, which had been suppressed in 1793, after its reestablishment.

All of this correspondence to Hachette and Mathieu reveals that Dupin, from entering the Génie maritime, was doing his best to attract the attention of the great scientists of the time (Monge, Delambre, Prony) and to establish friendships with them. This will be further demonstrated in his letters to Mathieu from Corfu. Throughout that correspondence, Dupin talked of papers to show the men for criticism and judgment, though he remained preoccupied with mathematics and applications. He frequently returned to his theories on cuttings and embankments and lines of curvature. It was in Corfu that he would begin to consider naval applications, with his 'Expériences sur la flexibilité des bois' (Experiments on the flexibility of wood; 1811), for example.

4.5. France, After Abandoning the Invasion of England Until the Treaty of Tilsit and the Milan Decrees

When Dupin was leaving Antwerp for the port of Genoa, what was the political situation of France? It had just suffered defeat at Trafalgar (on 21 October 1805), but it was beginning a successful two-month campaign with a victory against Austria at Ulm (in October 1805). Ulm was followed by Austerlitz (on 2 December 1805), marking the Austro-Russian defeat. The Russians retreated to Poland, and Austria signed the Treaty of Presbourg (on 26 December 1805). The following year, 1806, saw victories for France at Jena and Auerstadt (on 14 October), crowning the Saxony campaign. These were closely followed by the Polish campaign, with victories at Eylau (on 8 February 1807) and Friedland (in June 1807) against the Russians. This was also the year of the declaration of the continental blockade by the Decree of Berlin (on 21 November 1806) in response to the blockade of the French coasts that was instituted by England in May. The Decree of Berlin stipulated, in particular, that the British Isles were blockaded and that all trade and contact with Great Britain were forbidden.

The year 1807 was marked, a few days after the battle of Friedland, by a two-day meeting between the Russians and French which led to a treaty. On a large raft in the middle of the River Niemen, Napoleon and Alexander made their first contact on 25 June. Napoleon's demands of Alexander were for the alliance of France and Russia against England and the cessation of trade between Russia and England. The negotiations culminated in the signature of the Treaty of Tilsit on 7 July 1807 by Charles-Maurice de Talleyrand and the Russian representatives. It confirmed peace between France and Russia as well as Russia's ceding to France the Bay of Kotor and the seven Ionian Islands. The advantages of this treaty to France were enormous, and Napoleon was at the height of his power.

Two Milan decrees, the first on 23 November 1807 and the second on 17 December 1807, ordered, respectively, the seizure of any craft which had visited a British port and the seizure of any vessel which had agreed to submit to British law. The Treaty of Tilsit, followed by these two decrees, authorized Napoleon to apply the conditions of the continental blockade, to close British access to the northern seas, and to entrench the French position in the south through the fortification of Corfu and the Bay of Kotor. Thus it was that French engineers of the Génie maritime were transferred to these new French possessions. This was the case for Charles Dupin as well as for some of the *polytechniciens* who belonged to other services such as the Génie militaire, Dupin's friend Augoyat (see chapter 3.1) among them.

Endnotes

1. Admiral John Thomas Duckworth (1747–1817), perhaps best known for the relief of St. Domingo in 1806.
2. J. B. Marestier (1781–1832) in 1822 went on an 'observation' mission to the United States, where he studied steamships and military ports.
3. J. M. Tupinier (1780–1850) entered the Ecole centrale des travaux publics in December 1794. He was not yet fifteen years old. He was a colleague of G. Dufaud, one of the founders of the iron industry at Fourchambault. In 1823, Tupinier was the director of ports and arsenals; under Louis-Philippe, he would become the minister for the navy.
4. The work of the préfets maritimes was essentially administrative; they were not necessarily attached to the navy, and among them there were some generals. Later, in 1834, Dupin was to criticize them for their incompetence, as will be seen.

Chapter 5

To Corfu

5.1. Why Corfu?

Six months after the signature of the Treaty of Tilsit, on 20 August 1807 to be precise, the Bay of Kotor and the seven Ionian Islands became French. Napoleon had entrusted to General César Berthier the task of taking possession of Corfu. Berthier remained there as governor for a year, but in 1808, he was replaced by General François-Xavier Donzelot, who served as the governor until the French again lost the islands in 1814. It was one thing to take Corfu, quite another to supply it and defend it against the British, who were trying to blockade the entire archipelago. Admiral Honoré-Joseph-Antoine Ganteaume (1755–1818), of whom more will be said, was assigned this task, and he reportedly fulfilled his mission

> with his usual good humour and skill. Thanks to him, Corfu received an adequate garrison and food and munitions in such great quantity that the governor, General Donzelot, could resist any British attack and keep the tricolor flying over the Ionian

> fortress until the return to Paris of Louis XVIII in 1814. (SMM, *Le musée citadin*, documents 9, 12, 14, and 16)

Christophe-Augustin Lamare-Picquot's contemporary description provides the background to the conditions the French encountered:

> When Corfu was taken, we found the stores almost empty: little powder or ammunition. In February, 1808, the Emperor gave the order to Admiral Ganteaume … to trick the British cruisers with a false trail and to go and throw into Corfu supplies, powder, ammunition, lead and iron. The squadron remained in Corfu just long enough to unload all that we needed, and returned to Toulon without a shot being exchanged with the British. (Pernot 1918, 23)

The Corfu naval expedition was a result of the emperor's plans for Sicily and the linking of that island with the kingdom of Naples in the interests of his brother Joseph, the king of Naples. Napoleon was awaiting the surrender of Scilla, neighbouring on the Italian Messina Strait, which was occupied by the British. He aimed to accomplish both goals with the French squadron under Honoré Ganteaume, which was crossing the Messina Strait on its way to Corfu. However, because of foul weather, all did not go according to plan, and after suffering considerable losses of men and ships during the journey, when Ganteaume arrived at Corfu on 23 February with a sorely depleted squadron, he considered it wise to depart for France just two days later. The French fleet finally returned to Corfu on 15 March.

What had happened to Dupin, who had left Toulon with the French fleet on 10 February? Very little is recorded of his role during this time, but it is known that he was of great service to Ganteaume. It was Dupin and his team who repaired Ganteaume's ship *Commerce de Paris* and on 16 March returned it to the admiral. Dupin remained on the island when Ganteaume departed in March. Fox described this period of Dupin's life:

> It was Ganteaume's fleet which brought Dupin to Corfu in March 1808. He came, it seems, with freshly acquired skills in Latin,

> Greek and Italian, a burning antiquarian interest in France's Italian possessions, and a youthfully romantic Hellenism that responded readily to the beauties of the place and the degradation of the people. The pressures of his first duties, which included the organization of some rapid repairs to the French ships before their return to Toulon, quickly subsided, and he then found time for scientific pursuits, ranging from his own continuing research in descriptive geometry to experiments on the strength of materials which he conducted on masts in the naval shipyards, and for the advancement of Greek literature and national spirit. (1985, 306)

Honoré Ganteaume, born in 1755, had gone to sea at the age of fourteen and participated in the American War of Independence. A lieutenant in 1781, he was with Admiral Pierre-André de Suffren's (1729–1788) squadron in India at the siege of Trincomalee. In 1783, he became the captain of a fire ship, then served on *Apollon* before returning to France. For six years he served on vessels of the Compagnie des Indes. Taken by the British in 1793, then released, he returned to the navy as a lieutenant and became a ship's captain in 1794. He was in the Atlantic with Admiral Pierre Charles Jean-Baptiste Villeneuve and was then posted in Brest to protect the arrival of naval convoys. He was promoted to chef d'etat-major for Bonaparte's Egyptian expedition in 1798 and sent the news of the Aboukir defeat to Bonaparte. As contre-amiral, he remained in Egypt in charge of the naval forces on the banks of the Nile. He earned the favours of the first consul, became conseiller d'état, was promoted to the command of the Mediterranean fleet, and then was made préfet maritime at Toulon. Three times the minister for the navy in the absence of Decrès (1761–1820; see chapter 5.1), he commanded the navy from 1808 through 1810 in the Mediterranean. It was then that he was involved in the expedition to the seven islands.[1]

As for the naval engineers, a former naval officer and historian, Claude Farrère, declared, 'We had the best engineers in the world … Jacques Noël Sané … brought the art of building sailing ships to its highest perfection.' (Farrère 1934). J.-N. Sané (1740–1831) was acknowledged as the designer of some of the finest ships that were then afloat, and as a member of the first class of the Institut, he examined several papers that were presented by Dupin to that body.

5.2. DUPIN IN CORFU (1808–1811)

During a stay in Corfu of almost four years (from 23 February 1808 to near the end of 1811), Dupin was involved in a great variety of activities as an abstract researcher, a naval construction engineer, an educator (through the creation of an academy), and a man of the world (see Perrin 1983, 244–268). When he arrived on the island, he was twenty-four years old.

In mathematics, he went on with the work in geometry he had begun in 1805 in Holland and Italy and continued at Genoa in 1806 and in Toulon in 1807. He gave an account of this work in his article 'Analyse de plusieurs mémoires de géométrie', 'dans les momens de loisir que lui laisse son service' (in the few leisure moments his work allows) (*Correspondance sur l'Ecole polytechnique*, 2:387–396). Dupin dealt with questions of curvature, conjugate tangents, and the notion of the *indicatrix* (see chapter 3.5). He explained that his research had been very difficult because of the absence of basic documents in Corfu. He also studied the theory of orthogonal trajectory surfaces and its application to the determination of lines of curvature—work which earned him the title of foreign associate of the Institut royal de Naples in 1809. These various mathematical researches are included in Dupin's *Développements de géométrie*, which was published in 1813.

As an engineer, Dupin was involved with material and practical matters. He arrived on the island with 115 military workers. He wanted to remain there because there was much to be done, and he was responsible for the good running of the working sites, as is noted in this letter of 28 February 1811, addressed to 'M. Baron, *commandant supérieur de la Marine*':

> My commander
> According to your orders, I have taken precise details of our resources in sail canvas, to see if we can meet the demand for

> 14,000 [illegible] for the land engineers. It is quite impossible. In fact, our sail canvas is part of the new sails left by the Russians[2] and which, for the good of the service, cannot be completely sacrificed. The rest is falling to pieces and is in no state for any purpose. (ADN; PJ 20)

In a quite different area, Dupin, during 1811, his last year on the island, was experimenting in the Corfu arsenal on the flexibility and elasticity of wood, work which he was to take up again on a greater scale at Toulon and Dunkirk. The experiments were to be reproduced by the British during one of his visits to Great Britain. According to his custom, and as ever much concerned with attracting attention and appreciation, he addressed a paper on this subject, 'Sur la flexibilité du bois' (On the flexibility of wood), to the first class of the Institut (Academy of Sciences). It was to be published with acclaim in the *Journal de l'Ecole polytechnique* (1815, vol. 10, *cahier* 17). The paper was presented to the Institut on 12 July 1813, quite some time after Dupin's departure from Corfu; the examining committee of Carnot, Prony, and Sané gave it a favourable report.

Dupin began his paper with general remarks in which he indicated that the idea of applying mathematical science to determine the strength of wood had started with Galileo. Dupin's experiments were of a high scientific order, whereas previously, the use of timber in the navy had been somewhat empirical. Edouard Estaunié (1862–1942), a former *polytechnicien*, claimed (in his 1932 preface to *L'Ecole polytechnique*) that around 1800, 'and for many years more, the role of the naval engineer was above all, the judicious selection of wood, to use the best part and assemble it'. This comment holds cheap Dupin's experiments on wood and his subsequent research on the stability of floating objects, ships in particular (Dupin 1813a), which was conducted during many years after 1800.

Ever eager to promote himself, on 12 November 1813 Dupin wrote to his elder brother from Toulon, where he was undergoing a long convalescence after an illness he had contracted in Corfu:

> A little note for the printer, M. Greffier … I must have 100 copies of the analysis of my work on wood, as this seems to have been

> rather well received by savants and the Ministry. It will be more widely read by engineers than less easy and more abstract theories. I recently sent the same work to Poisson who replied that he had immediately sent it to the *imprimerie impériale* [imperial press] to be printed in the *Journal de l'Ecole polytechnique*, where the greatest savants enjoy the honour of seeing their publications. (DRB; PJ 21)

In his 'Expériences sur la flexibilité, la force et l'élasticité des bois', Dupin provided theories on the wood that was used in ship building, returning to curvature and the radii of curvature. He recalled the fine work of Girard[3] on the strength of wood, but stated that account had never been taken of the deformation that was produced in timber over time:

> When we construct our buildings, our machines, our ships, we assume that parts of large dimension and in addition less seriously stressed, retain the form intended by rigorous design: this is not the case … The biggest building that we can construct in wood is, without doubt, a ship … Hardly is it afloat on the sea, than the opposition of forces begins, on the one hand produced by the considerable mass of its extremities, and on the other, by the repulsion of the water, mainly concentrated towards midships; such opposition bends the whole of this great machine at the same time, causing it to curve, an 'arc' we have seen, over a length of sixty metres, sometimes of half a metre of curvature and more. (Dupin 1813a)

A twenty-first-century observation of the difficulties this impact can cause has been recorded:

> On this crossing, with waves at times averaging between 35ft and 50ft, QE2 was put under tremendous stress. The ship would plunge into a deep swell, shutter, rise sharply and wobble then dive again. Standing on the Boat Deck, one could look down into the deep troughs as they passed. In a quartering sea, a view along the length of Three Deck showed that the ship would actually twist. That's flexing.

> Pipes burst around the ship, in passenger corridors, in the deck head. (Scull 2007, 21)

Whereas studies of timber had been essentially concerned with resistance to rupturing under considerable stress, Dupin proposed to determine the comparable resistance of various substances when they were submitted to minor forces. These substances would not suffer greatly from the applied forces, and the form of the sample submitted to them would deteriorate but little; thus, what Dupin called, in his paper, their 'virtual resistance' could be found. The availability of timber had been a constant concern for the French navy. When Pierre-Alexandre-Laurent Forfait and Daniel Lescallier visited Britain on behalf of the French navy in 1789, they concentrated deeply on the treatment of different types of timber and gave a full account in their subsequent report of all that they had seen (Bradley 2005, 131–206). This was many years before Dupin was to visit Britain. The interest in British wood may well have grown after Trafalgar. In spite of the exposure of the fleet to enemy fire, few British ships were holed below the waterline—a tribute to the strength of the hulls, British oak, and British shipbuilders.

Describing his experiments in Corfu, Dupin explained that he had cut parallelepipeds of the same length (a little more than two metres) in oak, cypress, beech, and pine. Placing the various samples on two supports at their extremities, he loaded them in the middle with weights of increasing value. He studied the bending at the midpoint of each wood piece under the effect of the various loads. This led him to two theorems that were confirmed by his subsequent experiments:

> 1. The bending produced by very small loads is proportional to those loads.
> 2. When the same piece of wood is loaded between the same supports by different weights, these weights are reciprocally (that is inversely) proportional to the radius of curve of the piece at its mid-point, and the curve itself is, consequently, proportional to those very small loads. (Dupin 1813a)

Dupin explained that the samples of oak and pine came from the Russian ship *Michael*, which he had demolished in Corfu in 1810, the year before he began his experiments. In addition, he noted, when he conducted the experiment, he was in a port where everything that was needed to work with optimum precision, even perfect scales, was lacking. However, he added, the small differences between calculations and observations were always within the admissible limit.

5.3. THE BACKGROUND TO THE CREATION OF THE IONIAN ACADEMY

From the end of the eighth century BC, the island of Corfu had been a Corinthian colony; the island's request for help from Athens against Corinth had provoked the Peloponnesian War. During the Christian era, Corfu belonged in turn to Byzantium, Venice, and then to the kingdom of Naples (in the thirteenth to fourteenth centuries). It was ceded to France in 1797 and became involved in European affairs. In 1799, it was taken from the French by a combined force of Turks, Albanians, and Russians, and in 1800, it became an independent republic under Turkish and Russian protection. After the Russians left the islands in 1801, disintegration began. In Corfu, according to Fox (1985), 'there was a wave of rebellion and assassination which culminated in a victory for the democratic faction, the establishment of a new constitution, and the abolition of the privileges and titles of the nobility'. Internal strife continued for the next six years, until Russian intervention resulted in a dependency which lasted until Tilsit, when Corfu was ceded by Russia to France together with the other Ionian Islands. Corfu was a French possession from 1807 to 1815, when it became British. It remained so until 1864, when Great Britain renounced its protectorate of the islands, which then became Greek. In the twentieth century, Corfu was occupied by the Italians in 1941, and two years later, after Italy's surrender, it became a battlefield between Italians and their former German allies.

Fox continued and maintained,

> There is no evidence that the French administration which took possession of the islands in 1807 had any clear political or social mission … for him [Napoleon], the islands were of purely strategic importance and … only Corfu, the 'key to the Adriatic', as he called the island, was of real consequence. (Fox 1985, 305)

5.4. Dupin and the Academy

Dupin left his mark in Corfu, not least as one of the founders of the Ionian Academy, of which he became the secretary. He favoured public, free lectures in the most useful disciplines. Five courses were taught by the members of the academy, and Dupin assumed the chair of mechanics and physics. His work in that capacity presaged the courses he would teach and his contribution at the Paris Conservatoire des arts et métiers, as will be seen in chapter 12.

The new academy was based on the Institut d'Egypte, established in Cairo some ten years previously, in which Dupin's guide and mentor Monge had played a major role. In both institutions, cultural paternalism was highly apparent. Indeed, in his inaugural address to the Ionian Academy on 17 July 1808, Dupin spoke of the 'paternal hand' of France and the virtues of Ancient Greece as well as its weaknesses:

> We have just seen the causes of the rise and fall of the Hellenes! We have seen the dreadful lessons of experience traced in blood! … Follow the path provided for you, Ionians, and soon you will see your country, which flourished in the past, bloom again. Take a step to progress and you will see the French government smile on your efforts and make them fruitful … The government knows that rich, powerful and, above all, independent islands and islands without a navy were never the same. Your ports will be restored and maintained; new ones will be founded. (Dupin 1825a, 1:5–22)

The islands were in an appalling condition (see, for example, Bosset 1821). Dupin laid the blame for this on the Venetian Republic, taking a political stance in favour of the Turks. They had at least allowed the Greek language to be learned and used on the islands, but Dupin probably felt it was expedient to make the comparison with the Venetians, because Turkey was so close and a constant threat. Ties with the Greek mainland were now weak, and the Greek language was used neither among the educated classes nor in official transactions (Dupin 1809). Education was almost nonexistent, but with the creation of the academy, it was hoped that the ordinary young men of Corfu and the other islands would benefit from the free public lectures it would offer.

In the journal *L'hellénisme contemporain* (July–August 1947), in an article entitled 'La bibliothèque publique de Corfou' (The Corfu public library), Constantin Soldatos, the director of that library, briefly reviewed the intellectual activity of the island during the seventeenth and eighteenth centuries. Then he came to the first French occupation:

> But soon the armies of Bonaparte … arrived in Corfu, bringing ideas of freedom, equality and fraternity … Free social development began for the masses so long oppressed. The valuables of certain Catholic convents … were seized and claimed for public ownership. The library then became the Public Library of Corfu with some 4,000 books. Officers and French officials rushed to offer books …
>
> Some years later, the Imperial forces occupied Corfu. Despite the sudden change and the general instability, intellectual life in Corfu experienced a new departure with the creation of the Ionian Academy, whose aim was the advancement of the arts and sciences, and the development of the islands' productive forces. The Academy held sessions during which papers of the members were read; it awarded prizes, organized courses and conferences. The library was annexed to it and its administration was entrusted to a commission of academics from the Ionian Academy.

He made no mention of Dupin.

5.5. The Work of the Ionian Academy

The Ionian Academy was organized very quickly, and it was just four months after Dupin's arrival in Corfu that the inauguration took place. Few records of its work have survived (Fox 1985, 309), but in the first year, 1808 through 1809, there were courses in physics, chemistry, natural history, physiology, hygiene, anatomy, and surgery. Greek literature was added during the following year, and later, a course in jurisprudence was also introduced. One plan, briefly implemented in 1813 and 1814, was for thirty young islanders to be sent to France for technical education at the *écoles d'arts et metiers* (schools of applied arts and crafts).

The academy's teachers were French officers and local men, most of them doctors. Dupin's great friend Augoyat joined him in the running of the academy. The doctor and writer Nicolas Mavromati organized the course on Greek literature. He had received his medical training in Italy, having fled the Turkish rule in mainland Greece, and his experiences enhanced his successful lectures. The shortage of instruments and materials did not make it easy for the teachers, whose work, in the main, was probably as competent as could be expected.

Lamare-Picquot recorded the opening of the academy:

> The Ionian Academy has promised that professors chosen from amongst its members, would give public and completely free lectures on the principles of the sciences most useful to Ionia. Today, the Academy keeps its word; it opens its doors to all friends of knowledge, with elementary courses in physics, botany, physiology, political economy and criminal law. The aim of these courses is less to produce old-style savants than to offer to the greatest possible number easy learning of prime usefulness; they are courses stripped, as it were, of the luxurious aspects of science, and reduced to what can be presented simply and be immediately advantageous in its applications ... The courses will be fairly elementary and easily followed by the young people ... The Academy congratulates itself on seeing young mothers, or those who soon will be, coming to appreciate this learning which they will

> then pass on to their sons … M. Dupin will teach every Monday at midday, physics and general chemistry, with their most useful applications to be established in the Seven Islands. (Pernot 1918, 218–220)

He continued with a list of professors and their timetables as well as the details of prizes to be offered for the best work.

However, it soon became apparent that the audiences at the academy were not what Dupin had expected and hoped for. The lecturers found themselves talking to their colleagues rather than to eager young islanders. Dupin was aware of this after the first year of teaching and recognized it in his *Discours et leçons* (Lectures and lessons; 1825c, 1:2–3) as the main failing of the academy, which was also much dependent on his own personal involvement. He was up against the traditional values of the Greco-Venetian nobility, who were hostile to an institution where social status was disregarded and which they saw as a dangerous source of revolutionary ideas.

This was all a far cry from Dupin's original aspirations. In a dedication to Augoyat of his *Du rétablissement de l'Académie de Marine* (On the reestablishment of the Naval Academy) (see chapter 6.6), he expressed regret that their joint efforts had been 'useless'. Dupin had set off with such high hopes, as can be seen from the masterly piece of self-publicity he had produced when the Ionian Academy was in its early stages. On 7 April 1810, he had written to Mathieu, 'Herewith an idea of the academic volume I'm going to put together and send to M. Lacépède,[4] so he may present it in homage to the *Institut*.' It was to be a collective work, and Dupin then listed various articles which would appear in the book:

1. History of the Ionian Academy during its first year, by Dupin.
1b. Report by M. Massapoulo on M. Dupin's papers on the teaching and study of law.
2. Comparison of French weights and measures with those of the seven islands and Albania, by Augoyat, Dupin and Arnaud, former scholars of the *Ecole polytechnique*.

3. On the establishment of vaccination in the seven islands by Dr Gangadi of the Italian Academies and the Georgophiles of Florence.[5]
4. On the treatment of quartan ague [malaria], by Dr Franck of the *Commission des sciences et des arts d'Egypte*.
5. Augoyat on the artificial grindstones used in Greece and the Ionian islands.
6. On the fine arts by M. Deluret, military engineer[6] [on the work of Deluret, see Montagne 1979].
7. In praise of the modern Greek Boulzaris who translated the whole of Vergil into Homeric Greek verse, by Dr Marromatis. Eulogy written in ancient Greek, translated into French by Dupin.
8. Hymn to Emperor Napoleon in Pindaric Greek verse by the same.
9. Italian ode on the crossing of the Danube and the battle of Wagram, by M. Delviniotti, of the Italian Academy.
10. Translation of the first song of the Greek poet Quintus Smirneus, continuator of Homer, by the same Delviniotti.
11. Translation into vernacular Greek of the conspiracy of Catilina de Salluste, by M. Capo d'Istria.[7]
12. Note on two fine collections of Greek and Roman medals existing in Corfu and containing several valuable medals and other new ones, by M. Pro Salendi, well-known antiquarian of [illegible] who intends to work extensively on these collections. (ADN; PJ 22)

Was it really possible that Dupin was able to translate from Ancient Greek after such a short time? Fox spoke of his 'freshly acquired skills' in the language (see chapter 5.1), and in 1810, Dupin produced his *Essais sur Démosthène et sur son eloquence* (Essays on Demosthenes and on his eloquence), in which he translated the Olynthian orations of Demosthenes. To this he added *Considérations sur l'éloquence de l'auteur athénien* (Thoughts on the eloquence of the Athenian author), which won the acclaim of the distinguished Hellenist Paul-Louis Courrier

(1837, vii–viii). Although it was published in 1814, the preface is dated 'îles ioniennes 1810'. A review also appeared in England in October 1814 by Charles James Blomfield, who was to become the bishop of London (Blomfield 1814). Dupin's programme sounded impressive and ambitious, but Lamare-Picquot commented, 'During our occupation, Captain Charles Dupin made great efforts to establish a sort of *école normale* to produce capable teachers. All he achieved was to reveal the complete absence of any existing means to propagate elementary education' (Pernot 1918, 44).

When the French were ousted from the Ionian Islands in 1815, the academy lost its legal existence. However, the groundwork had been laid, and the British who succeeded the French granted, in 1817, a charter for the creation of an academy for the various branches of the sciences, literature, and fine arts. Capo d'Istria, one of Dupin's collaborators in the production of the *volume académique* (academic volume) that was prepared by Dupin and destined for the Institut de France, was also one of the promoters of this new academy, which opened on 21 May 1824. However, according to Fox, under British administration,

> the Academy seems to have assumed a purely symbolic function as a means of bestowing distinction on local citizens and of recognizing, through election to corresponding membership, the islands' cultivated friends and benefactors, nearly all of them in France and Italy. (1985, 310)

Sadly, the records of the French period at the Ionian Academy have disappeared.

One of Dupin's students was the mathematician Jean Carandinos, who is considered to have launched the study of modern mathematics in Greece 'as much through his enlightened teaching as his important translations at the beginning of the 19th century' (Phili 2006, 83). The determination and talent of the young Carandinos must have impressed Dupin; under Dupin's valuable guidance, Carandinos studied mathematics in private and made great progress in higher analysis and mechanics before embarking on a career in teaching and research (Phili 2006, 84).

5.6. Dupin Leaves Corfu

Just over three years after his arrival in Corfu, Dupin was obliged to leave the island for health reasons. His correspondence provides details of his illness and the vicissitudes he and his companions had suffered. The first extract of a letter, written during his return journey and dated from Pisa (5 October 1812), is addressed to Jacques-Antoine Forestier, the head of the first division of the Ministry for the Navy and the Colonies:

> You are aware of my misfortunes; there were four of us when we left Corfu, my three travelling companions died within the space of a fortnight; and, after 15 months of suffering, I am still in a poor state of health. (ASHM CC7 779; PJ 23)

Dupin proceeded to request leave, which was granted by Sané (see chapter 5.2), the inspector general of the Génie.

The second letter predates the previous one, but provides additional information on the illness. It was addressed to Dupin's father from Naples on 27 December 1811, some five months after Dupin's departure from Corfu:

> My dear papa,
>
> Finally the spell is lifted, and for the first time for so many months, I can receive a letter from you … The kinkina [*sic*] had the effects that you feared for me; I have obstructions, but they are not painful and they are gradually yielding to the medicines I am using to combat them (DRB; PJ 24)

The reference to quinquina indicates that Dupin was suffering from malaria, which was very common in the Mediterranean. Quinine, an extract from the bark of the quinquina tree, was discovered in 1820 by Pierre-Joseph Pelletier and Joseph-Bienaimé Caventou, but before that, the bark itself was used as treatment. Dupin continued,

> What! If I know enough Italian to make myself understood? Don't you know that I lived for five years among Italians and very frequently in purely Italian circles, and that we almost always

> spoke and wrote in Italian in the Ionian Academy? and that I'm somewhat high-powered in Italian literature? Hang it all! A man who has had seven or eight Italian sonnets composed in his honour! I can't cope with this any longer and I shall have to acquaint you with all my renown.[8]

He returned to the subject of his treatment or, more precisely, to the doctor who was treating him:

> Fortune having at last tired of persecuting me provides me, on my arrival in Naples, with a doctor who does not cure his patients, but sings them … Having dispatched a wife, he produces a funeral oration in elegiac verse. He repeats how the potions, sedatives, stimulants, rhubarb and senna, nothing could have prevented her from drinking the fatal waters of the Lethe![9] He repeats with what patience she took the bitterest medicines and her virtues and her faith in the unfortunate doctor. Then he weeps over his verses with the poor husband and they gradually console each other in the comforts of Hippocrates and the Muses! (DRB; PJ 24)

He added that he intended to leave immediately for Rome and to stay there until the end of February 1812.

In 1813, Vice-Admiral Valentin de Lessegue suggested that Dupin should return to Corfu, but Sané opposed this in a letter dated 19 February 1813: 'The request … could not be granted, for Dupin has just been appointed to the command of a company of military workers in Toulon … The military workers in post in Corfu, because of their ill-discipline, should be replaced' (ASHM CC7 779; PJ 25). Dupin was never to return to Corfu.

Endnotes

1. The seven islands are Corfu, Paxos, Cephalonia, Ithaca, Zakinthos (or Zante), Kithira, and Levkas.
2. In 1800, the islands had been declared an independent republic, the Septinsular Republic, under Russian and Turkish protection. The Russians forces left the islands in 1801.
3. Pierre-Simon Girard (1765–1836), a *ponts et chaussées* engineer well known for his work on canal development and one of Prony's closest friends and collaborators. Prony considered Girard to be an engineer with a well-deserved reputation for science and talent (see Bradley 1998, 15 and 24).
4. Comte de B. G. E. Lacépède (1756–1825), a French naturalist and politician. He became a minister of state in 1804.
5. The Academia dei Georgofili, founded in Florence on 4 June 1753 for the development of agrarian science and economics—work which continues there today.
6. Deluret was a captain in the Génie militaire. The military records describe him as Deluret *aîné*. In 1810, he was with the French army in Italy.
7. Capo d'Istria, born in Corfu in 1776, devoted himself to the cause of Greek independence.
8. Dupin used the word *gloire*, a French concept that is notoriously difficult to translate.
9. The Lethe is one of the several rivers of Hades. Those people who drink from it fall into oblivion.

CHAPTER 6

DUPIN'S RETURN TO TOULON

6.1. CONVALESCENCE

Dupin's mental state during the period between his hurried departure from Corfu in July 1811 and his resumption of duties at Toulon in February 1813 was marked by some depression and also by an urge to escape from the professional routine. He even considered quitting the Corps du génie maritime. The following letters, several of which came from his father, indicate the evolution in Dupin's thinking. They also illustrate the constantly aggressive attitude of the father toward his second son.

On 26 January 1812, Dupin père wrote to his eldest son,

> I have received a letter from Charles and I am not pleased with it. His letters, extremely rare, and very brief, speak only of literature and twaddle; too little on his state of health and way of life, nothing at all concerning his profession as an engineer. It could be said he has resigned his post or that he has no fear of being considered as having resigned. (DRB; PJ 26)

Dupin's father was not far wrong, as shown by the following letter Charles wrote to him:

> Naples, 19 March 1812
> I am applying for the post of secretary for maritime construction, in Paris, a position only available to a *sous-ingénieur*, and which would not relieve me of my post!
> This job is perfectly suited to my taste for study. What is more, I feel so deeply my need to live in Paris, so as to reap the fruits of my work and my acquaintances that if, in a year, I don't obtain some post that fixes me there, I shall submit my resignation pure and simple and I'll go and live in Paris on my small savings until I find employment. But how much more advantageous it would be to remain in my *Corps*, to have a job where one could develop one's talents and live close to brothers and family.
> Recommend me, therefore, to your old friends so that they write in my favour, either to the Minister or to M. de Sané, or to Vice-Admiral Rosily, president of the *Conseil de constructions*.
> Farewell, my dear papa, reply to me at Rome, where I intend to go within a fortnight, my health being somewhat improved. (DRB; PJ 27)

Even in his state of depression, Dupin was conscious of the importance and usefulness of having friends in high places, but as will be seen in future chapters, the idea that he would be able to live on his 'small savings' is hardly credible. He was not a man to budget wisely.

On 2 April 1812, Dupin's father responded to his son's request by writing to his eldest son, whom he asked to confer with Lazare Carnot (who was his close friend) and with de Sané on the possibilities or, if there were none, to give good reasons to 'our young man' (DRB; PJ 28). In a word, this was to evidence the 'zeal and devotion' of his contacts. Dupin père then advised his eldest son that it would be easier to reach Rosily 'through one of my old acquaintances'. The condescending attitude of Charles Dupin's father towards him is discernible in this letter; it would seem that there was constant division in the family. The youngest son, Philippe, wrote a note to Dupin *aîné* in June 1812 to the effect that their mother was always upbraiding them all, one after the other: 'It is only Charles who receives her blessings, because, knowing nothing about her affairs, he can't give an opinion' (DRB).

However, the father sometimes softened towards Charles. On 15 June 1812, he wrote to his eldest son that he had received word from Charles 'under the tower of Pisa'. He continued that he would have preferred that Charles be 'under the Saint Martin tower' (probably the twelfth-century bell tower of the Eglise Saint Martin), but that his health was improving and he hoped that after another period of quarantine—'may God will the last'—his son Charles would take off again (DRB; PJ 29).

Where Charles went to stay next is not clear, but it must have been close to his elder brother, to whom their father wrote on 3 December 1812,

> I am very sorry that your brother is not staying with you.[1] At least, make sure that he has all that is necessary where he is and above all, a fire because, for a convalescent, chilling can cause serious relapses. Take care of him for all that does not concern him sufficiently, that is himself and his true interests. (DRB; PJ 30)

This concern is most unusual on the father's part. What he considered to be Dupin's 'true interests' is a matter for conjecture, but it is true that Dupin was only twenty-eight years old; in his poor state of health, he might have made a rash decision.

His father continued to express his worries to his eldest son. On 4 January 1813, he wrote,

> Charles never writes anything to me about what is relevant to him personally, nor about his health. What do the doctors say? I can think of nothing more indecorous than a savant who communicates nothing of what one would want to know unless he is questioned face to face. (DRB; PJ 31)

Charles Dupin's illness lingered for a long time—from the end of 1811 until early in 1813. After a rest cure in Pisa in 1812, he was still in poor health.

6.2. At Last to Toulon

In February 1813, Charles Dupin was sent to Toulon. In 1816, he would go to Dunkirk after spending a few months in Marseille starting at the end of 1814. His long period of convalescence and depression was soon to be left behind, followed by a rush of activity, the acquisition of honours, and the beginnings of fame.

In 1812, he produced an edition of Léopold Vaccà Berlinghieri's account of Caesar's victory at Alesia. The year 1813 was to be marked by the publication of his *Développements de géométrie* (see chapter 3.6) and the acceptance by the Institut of the 'Mémoire sur la flexibilité, la force et l'élasticite des bois' (to be published in 1815 in the *Journal de l'Ecole polytechnique*). In 1814, Dupin presented his 'Stabilité des corps flottants' (Stability of floating bodies) to the first class of the Institut, and it was very well received. He also completed the first part of his *Tableau de l'architecture navale* (Survey of naval architecture). The *Développements de géométrie*, also of this period, has been described in chapter 3.6, and 'Sur la flexibilité, la force et l'élasticité des bois' was described in chapter 5.2.

6.3. Honours and Family Matters

As regards honours, in March 1813 Dupin became correspondant de l'Institut and, in the same year, chevalier de la Légion d'honneur. Having learned of Charles's nomination to the Institut, Dupin père commented (on 20 March 1813) to his eldest son,

> I am most concerned that your mathematicomaniac has become a corresponding member of the *Institut* when he will never be, before all and above all, a corresponding member of his family. However, I have finally received a letter from him which implies another, apparently remaining in his jumble of papers and, to crown it all, full of his usual nonsense which is not at all what I would want to find in letters from my offspring … He is making his mathematical twaddle unbearable by talking about it. (DRB; PJ 32)

On 9 July, Dupin père wrote again to his eldest son,

> I have received the portrait of Charles. I have never seen a portrait which less resembles its subject … All of you, such as you are, have no idea how to get yourselves painted and you think you recognize yourselves in portraits as long as they have a nose and ears … I am seriously afraid that Charles will quit his post or be relieved of it for his rebelliousness, and that Philippe will be suspended from his. Thus the pair of them … are the root cause of my ever increasing stomach pains. (DRB; PJ 33)

The fear for Philippe was occasioned by the possibility of his conscription, but the father's main concern seemed to be for his own health.

With regard to Charles Dupin's attitude to his brother Philippe, on 26 October 1813, he wrote from Toulon to Dupin *aîné*,

> My dear Dupin,
>
> As for Philippe, I excuse him less and I don't complain of him, for he is still young. He will learn the value of friendship when he needs people. I spoke coldly of the *Domaine de Coeurs* to my father; I shall be equally cold on the subject of material interests … My dear Dupin, understanding and no ill-humour between us; you will never love me as much as I love you.
>
> Your brother,
>
> Charles Dupin (DRB; PJ 34)

The fine property, Domaine de Coeurs, a part of Mme. Dupin's dowry, was the object of litigation in the parents' divided household.

The extent to which Dupin depended on his elder brother is again illustrated in a letter he wrote from Toulon on 12 November 1813:

> I should have much preferred, I must admit, that you had seen the Minister for the Navy personally. I have never been able to ascertain whether you had received the letter of introduction that I had taken the precaution of sending you through the *Conseil de constructions*. Did you receive that letter? Did you attach it to my book? I have already completed the analysis of Carnot's work; I send it with my letter and ask you to convey it to that worthy

> man. It's the least you can do to thank him for his friendship and kindness to you and to me. (DRB)

The letter continues with family and personal details:

> Give me the address of a person in Auxerre to whom I can send the oranges ripening in my garden and other produce from Provence … What you tell me about *Messieurs* the Parisians is a great embarrassment to me, as I have 2,000 F to send to you and I don't know how to do it safely. (DRB; PJ 35)

This was the repayment of a loan to Charles from his elder brother. Charles was often short of money, and his brother frequently came to his assistance. Dupin continued,

> You'd be very kind if you saw Mme Courcier a little. 20 pages on Demosthenes have been printed in six months.
>
> P.S. If I go to the *Grande Armée*, I'll do the impossible to come through Paris and hold you again to my heart.
>
> If you find my letter too expensive, charge it to my account. (DRB; PJ 35)

At that time, it was the recipient of a letter who paid the postage; the first French postage stamp appeared in 1849. The concern over conscription was very real. Napoleon had created an army of five hundred thousand men, with one hundred and forty thousand conscripted in 1813 and more to be found in 1814. It was no longer the Grande Armée of experienced soldiers, but one of young troops who would be receiving their training on the battlefield (see Tulard 1970, 272 and Bradley 1975).

Dupin wrote again to his brother on 23 November 1813 to the effect that he had received the printed versions of his last oration of Demosthenes and the beginning of his 'Considérations sur l'éloquence de ce grand Athénien' (Thoughts on the eloquence of this great Athenian). He added that he had made many changes thanks to the excellent advice of his brother (DRB; PJ 36).

6.4. Dupin's Work in Toulon

Such details reveal information about Dupin the man while he was working as an engineer in Toulon. On 8 September 1813, Contre-amiral Lhermite wrote to the minister for the navy and the colonies that Dupin, a sous ingénieur employed at Toulon, had completed the work he had begun in Corfu on the flexibility and strength of wood. He added that it was nothing less than Dupin's indefatigable dedication and his extreme devotion to his work that had enabled him to conduct such essential experiments. These were done independently of his duties, which he carried out in every respect with complete satisfaction. The head of the Génie depended completely on him, both for the construction of *Colosse* and for the work in the port connected with ship repair (ADN *fonds* 20 J 31).

Dupin would spend most of 1814 in Toulon, until the end of November. The preparation of the first part of his *Tableau de l'architecture navale* occupied a great deal of his time, as he explained on 3 August 1814 to the préfet maritime: 'At present, I seek only to make progress with my work on "naval architecture"; several engineers have had the goodness to provide me with superb drawings for it and everything is favourable for the completion of my undertaking' (ADN; PJ 31).

It will be seen that the work would be of great assistance to Dupin in furthering his career as well as in developing his plans for study visits—in particular, visits to Britain, when the political climate became favourable. Also in his letter of 3 August 1814, he mentioned other matters, explaining that during the course of describing the intricacies of rigging, he had become interested in nautical knots and the points at which the ropes cross. This could become an important aspect of what he called *géométrie de position* (geometry of position).[2]

6.5. The Toulon Naval Museum

The final months of Dupin's stay in Toulon saw the birth of the Toulon Naval Museum, which was so closely associated with his name. In his *Mémoires sur la Marine et les Ponts-et-Chaussées de France et d'Angleterre* (On the Navies and Bridges of France and England) of 1818, Dupin referred to the museum:

> Towards the end of 1813, I presented to Rear-Admiral l'Hermitte [*sic*] [Jean-Marthe-Adrien Lhermitte (1766–1826)], *préfet maritime* of the Toulon region, a description of several of Puget's[3] masterpieces deposited, and effectively abandoned in the Toulon naval sculpture workshop. To this description was attached a memoir where I proposed a few ways of preserving these precious works of art and of improving the study as well as the practice of naval sculpture. General l'Hermitte, to whom the Toulon arsenal owes the creation of several useful establishments, was quick to respond to the means of preservation I had elaborated … By a decree of 3 September 1814, the préfet decided:
>
> 1. That a single establishment should thenceforth constitute the depository for all valuable works of naval sculpture and architecture, left in cabinets or spread around the workshops of the Toulon arsenal.
> 2. That a site partly occupied by superfluous and ill-situated workshops should be given over to this establishment and its outbuildings.
> 3. That *capitaine du génie maritime* Charles Dupin should be entrusted with the planning and execution of the project.
>
> Some time later, I arranged that the Establishment should bear the name of *Musée maritime*; that a library should be added to it, whose site would be taken from the adjacent sculpture workshop; finally, that at the end of the same workshop, a school of naval sculpture and design would be established. (Dupin 1818b)

At the time of its creation, the Toulon museum was at the Arsenal Pavillon de l'horloge. Since 1981, it has stood on a specially designated site. As a result of its establishment, a similar museum was begun at the Rochefort arsenal by Jean-Baptiste Hubert,[4] also of the Génie maritime and director of naval construction at Rochefort from 1814, to display his own models and machines. Albert Chatelle, the deputy curator of the *Musée maritime*, made no mention of Dupin's role in the museum. After a brief description of Lhermitte, who was quick to rally to the cause of Louis XVIII upon Napoleon's departure for Elba in April 1814, he stated,

> Then, in September 1814, our *préfet maritime*, is expecting a princely visit. His enthusiasm for the new régime will finally endow the arsenal officially with a naval museum … it would be pleasant to show the great admiral and several other princes [*sic*] the models in the arsenal.
>
> This plan was gestating since year II, a time when the sculptor Félix Brun was asking the Convention, through a local committee for the arts and sciences, permission to dispose of articles related to the navy for the creation of a museum. What models were concerned, if not those so evocative of the great Navy of Louis XV and Louis XVI, without forgetting the magnificent sculptures of Louis XIV's galleons?
>
> In his agenda, the *amiral préfet* orders, on 2 September (1814) that the hall of models and sculptures, naval architecture etc., must be arranged in the most suitable manner and that the display of the models must be completed for 10 September at the latest'. (Chatelle 1943)

Dupin's work was not recognized, just as Soldatos gave him no credit for the creation of the Ionian Academy (see chapter 5.4).

The oversight is somewhat redressed in a typed manuscript in the Toulon Library by Commander Emanuel Davin, who wrote,

> The first efforts for the creation of a naval museum were those of a master sculptor, Félix Brun, who, in 1796, assembled some figure-heads, the work of Puget. This sculptor, spoke of it to Dupin, who proposed the creation of the museum to Rear-Admiral Lhermite[5] (1766–1826); he decided on its installation in the *Pavillon de l'Horloge* … on 3 September 1814.

He did not add that *capitaine du génie maritime*, Charles Dupin', as Dupin would have wished it, should be entrusted with the planning and execution of the project, so one must conclude that Dupin was inclined to exaggerate his role in the matters with which he was involved.

Nevertheless, his name now remains closely linked with the creation of the museum.

6.6. Dupin Leaves Toulon

At the end of November 1814, Dupin was to leave Toulon, being posted to Marseille to supervise the completion of the ships *Bonite* and *Hypopotame*. During his time in Toulon, he had finished his 'Stabilité des corps flottants', which was presented to the first class of the Institut de France on 10 January 1814. On 30 August of the same year, the academicians Jacques-Noel Sané, Louis Poinsot, and Carnot reported on the work, stating that Charles Dupin 'has found the secret of saying new and interesting things on a subject we thought was exhausted' (*Procès-verbaux de l'Académie des sciences*, 30 August 1814). Having first addressed the work of Euler and Pierre Bouguer, they added,

> M. Dupin follows a very different course from that of those two illustrious geometers. He uses a geometry unknown in their time and this new instrument leads him to new results … M. Dupin has theorems which embrace all those already known on the stability of floating bodies, and then has many new theorems … This new work from M. Dupin confirms the expectations resulting from his earlier ones, and one can only applaud his constant efforts to direct his results towards the practice of the fine art to which he is dedicated. We consider that Dupin's paper merits the approbation of the *Classe* and we are suggesting to him that it be included in the collection of the *savants étrangers*. (*Procès-verbaux de l'Académie des sciences*, 30 August 1814)

This was high praise indeed.

In 1815, Dupin was calling for the reestablishment of the Académie de marine and published a booklet to this effect, *Du rétablissement de l'Académie de marine* (On the reestablishment of the Naval academy). This academy had been founded in 1748 as a private institution to bring together mariners and savants (especially geographers and astronomers) who were interested in the development of the navy and in scientific problems related to navigation and naval construction. In 1793, it was closed along with all the other academies. In advocating its reestablishment, Charles Dupin opposed the idea of an Académie de marine centred completely in Paris, instead envisaging an academy that would be spread across various ports, with Paris as the focal point. As he put it,

> In order to advance the work of the *Académie de marine* towards a single goal, it is enough to create separate classes of academicians meeting in each large port, and to consider the Paris class as the nucleus, the centre for all the others. (Dupin 1815a)

Despite Dupin's efforts, which were united with those of other great mariners such as Pierre Forfait, the academy was overlooked by the authorities until it was finally reestablished in 1921.

Before following Dupin to his next posting at Dunkirk, I would like to recall his spirit and courage in advocating his democratic ideas and his vigorous defence of Lazare Carnot. After Napoleon's publication, in 1814 on his return from Elba, of the *Acte additionnel aux constitutions de l'Empire* (Additions to the constitutions of the Empire), Dupin wrote a paper demanding representative institutions and praising Carnot and Lanjuinais (Dupin 1814b). Then, after the news of the Waterloo defeat, when the White Terror was raging against the Bonapartists, Dupin produced 'Pompe funèbre à célébrer en l'honneur des guerriers français morts pour défendre la Patrie' (a funeral oration in honour of the French soldiers killed in their defence of the homeland). This was quoted in the *Dictionnaire des parlementaires français* (Dictionary of French parliamentarians), but it has not been possible to determine where the oration was originally published. In what was becoming his customary grandiloquent style, he called on the people of France: 'Let us raise our heads in the midst of danger … let us staunch the wound that still bleeds in our hearts, and fly back to the fight, if we cannot obtain peace with honour' (DPF).

Once more taking up Carnot's cause, Dupin protested strongly against the order for his proscription and prepared a defence. This, however, remained unpublished at the request of Carnot himself.

6.7. Dupin Returns to Dunkirk: The Decline in His Scientific Activity

In 1816, Dupin was in Dunkirk and eager to draw attention to his achievements as well as his future plans. On 18 March 1816, he wrote to Louis-Marie Bajot, the chef du Bureau des Lois au Ministère de la Marine,

> I am sending you a description of the Puget sculptures which was very well received by the *Institut*. If it is suitable, you could insert it into whichever of your editions you like. I shall then send you the description of the naval museum I have established in Toulon. Would you kindly ask M. Portier if, as he promised, he has submitted his report on the journey I wish to make to England. (ADN *fonds* Dupin 4J; PJ 37)

Such a journey was at last possible for Dupin. Waterloo was over, and the Channel coasts were clear; France was now under the second Restoration. The Congress of Vienna had done its work, and the 1815 treaties left France impoverished and smaller than it had been prior to the Revolution. As for Great Britain, she had established her predominance over her colonies and the seas. She had taken possession of the Ionian Islands. Combined with Malta and Gibraltar, this guaranteed her mastery of the Mediterranean. The country had taken Belgium from France and restored the balance in western Europe. As for Charles Dupin, he seems to have emerged unscathed from the general upheaval, having escaped even the conscription he feared in 1813.

However, in Dunkirk, he seemed to be rather less concerned with science and naval engineering in his eagerness to visit the British arsenals, and he began to be much more absorbed in economic issues. Moreover, he was increasingly interested in honours, particularly the title of academician. He was a correspondent of the Institut (first class) beginning in March 1813, but he had his eye on greater things, as is apparent in his letter to his old friend Mathieu, for example, written from Dunkirk on 3 April 1816: 'I am in complete ignorance of nominations for the title of *académicien libre*. Must one be a viscount, marquess, geometer, physicist, astronomer, I have no idea' (DRB; PJ 38).

To his elder brother, he wrote, also from Dunkirk on 6 April 1816,

> Eight vacancies for *academiciens libres* have been created at the *Académie des sciences* by the new organization.[6] I wish to be among the candidates for one of them. I have written to this effect to M. Delambre, the secretary. If you could see Messieurs Daguex and Lefèbre-Guneau to obtain information on the manner and the exact time of the election, so as to let me know right away, you would be doing me a great favour … My appointment to the Academy would change the face of my affairs and would certainly get me posted to Paris by the Minister for the Navy. (DRB; PJ 39)

In the spring of 1816, Dupin was already a correspondent of the Institut, a chevalier de la Légion d'honneur, and sous-ingénieur first class, with the rank of ship's lieutenant. He had won fame with his *Développements de géométrie*, which incorporated much of his mathematical work, and his 'Flexibilité des bois' and the 'Stabilité des corps flottants' (the first paper of part 2 of his 1813 *Développements de géométrie*) had placed him in the top ranks of the Génie maritime. He would have to wait for a seat in the academy, but in the meantime, he would concentrate all of his efforts towards his mission to Great Britain. He fully realized that the account of such a journey would stand him in good stead with the academy. He had become more the economist and educator of his future career, although he would remain a professional mariner. He would rise to the grades of director of naval construction in 1838 and inspector general of the Génie maritime in 1843. He had yet to publish his *Applications de géométrie*, which would appear in 1822, grouping the work he had carried out from 1814 to 1816. However, it could be said that the year 1816 marked the end of his purely scientific work and his actual engineering duties.

On 7 February 1816, the mathematician and physicist Fourier requested the minister for the navy to authorize Dupin's plan to observe construction procedures in other countries, particularly in England (ASHM CC7 779). Fourier attached Dupin's 'Plan d'un voyage en Angleterre' (Plan of a journey to England), which he had prepared in Paris in January 1816. Dupin asked the minister to sanction the following mission:

> To go to England to observe the state of the Arts directly or indirectly related to the Navy and to describe:
>
> 1. The agents and incentives used by the English for accomplishing their work in these arts.
> 2. The instruments and tools in use for producing correct forms and qualities.
> 3. The mechanical and geometric processes used to transform the raw material in its crude state to an industrial product.
>
> Then, to consider the application of these methods to the construction, rigging and arming of ships. (ASHM CC7 779)

Dupin pointed out to the minister that

> never was a study of this nature more important than that of the British Navy: never was the time more opportune. A long war, fertile in great expeditions, all with different aims, means and results, has created in the maritime arts previously unknown needs and unexpected resources to meet them … Thus, the lethargy of peace, which renders the art stagnant in our idle arsenals, will offer us the most advantageous means of improvement, for an impartial and reasoned comparison of our work with those we emulate … In France, there are two equally erroneous and pernicious viewpoints. Those of the first opinion consider our naval work to have attained the highest perfection … Those of the second believe, on the contrary, that everything is perfect in England… I do not believe in the perfection of our navy, still less in that of the English.
>
> Let us add a word about the means of accomplishing this. It is as a geometer and physicist that I shall travel England. It is with the recommendation of the *Institut de France*, several academies and British engineers. Our friends and those of my family will provide me with other facilities and, in spite of the shroud of mystery which will more than once impede my research, I am sure to see enough to acquire a fair idea

of the establishments I am unable to penetrate. (ASHM CC7 779; PJ 40)

Dupin ended his report by pointing out to the minister that it would be advantageous for him to leave as early as April to take advantage of the good weather for visiting English sites and workshops. He added that upon his return he intended to present a description of the establishments he visited; to describe the machines, instruments, and tools whose imitation or purchase would be useful to the French navy; and to keep for his *Tableau de l'architecture navale* the theoretical ideas to which his observations would necessarily lead him.

The young engineer was growing impatient. On 14 July, his father wrote to his eldest son,

> Give Charles all the necessary advice to keep well and behave well in England, so as not to expose himself to the Aliens' Bill or to Bedlam. Bogne [a cousin] claims that he will see nothing, as they show nothing to foreigners.
>
> Prevail upon him not to abuse the freedom of speech, so as not to be in trouble with the English police or the French police. I want him to come back with a head as cold as the glaciers of Switzerland and as measured in his conversation as Jacques Rosbif. Wish him a good journey for me. (DRB; PJ 41)

Finally, the time for departure arrived. Dupin wrote to Bajot,[7] the director of the Bureau des lois au Ministère de la Marine, on 21 July, 1816,

> Within a month, I will send to you a report that the *Institut* has just produced on my 'mémoires de physique mathématique' … I received my papers yesterday and I leave tomorrow … They have given me 1,200 as an allowance for six months and I am taking 3,000 for three months. (DRB; PJ 42)

On 21 July, he again wrote to Bajot, 'I have received my papers and I leave tomorrow'. On 22 July 1816, Dupin left for England.

ENDNOTES

1. Dupin *aîné* was then living at 221 rue Hautefeuille in Paris.
2. This 'analysis situs' had been considered by Gottfried Leibnitz and developed to a small extent by Leonhard Euler and Alexandre Vandermonde, but it assumed real importance in the nineteenth century, in the work of G. F. B. Riemann (1826–1866), the author of *Topology and Physics*, and H. Poincaré (1854–1912). The name today is 'topology'.
3. Pierre Puget (1622–1694), a naval sculptor.
4. Hubert, a successful designer of frigates and steamers, would become the director of naval construction at Rochefort from 1831 to 1845.
5. The spelling of this name varies; it appears as Lhermitte and Lhermite.
6. In 1816, the classe des sciences physiques et mathématiques of the Institut had again assumed its former title, the Académie des sciences.
7. L. M. Bajot published, with M. Poirré (the deputy director of the ministry bureau), the *Annales maritimes et coloniales* (Naval and colonial annals) in 1841.

CHAPTER 7

DUPIN'S FIRST STUDY VISIT TO THE BRITISH ISLES[1]

7.1. WHY TO ENGLAND?

Dupin's patience was finally rewarded, but he had had to wait for ten months for permission to go to Britain. This is perhaps not surprising because he had become suspect through his defence of Monge and Carnot. Eventually, he was able to embark for the first of four annual visits he was to make during the coming years. He left behind him a land where roads and bridges had been neglected or damaged, the merchant navy was disorganized, and the economy was stagnant. 'The greater part of the country's energy had gone into reform, revolution or conquest, rather than into industry' (Henderson 1966, 8; see also Robinson 1958, 3–15). Meanwhile, Britain had been forging ahead and was now far stronger financially, commercially, and industrially than it had been in 1793. With new colonies and overseas markets, the country had built canals and improved its road network. Above all, she had developed steam power on a large scale and was beginning to apply it to transport.

The war had inevitably disrupted trade, and there was a postwar depression in coal mining and iron making; food prices were also affected. In the main, however, England's manufacturing industry was prospering and her economy was resilient. France, in contrast, was weary and humiliated. Her progress was hampered by a shortage of coal and other essential raw materials in addition to restrictive guilds, heavy taxation, and bureaucratic limits on commerce and navigation. There was also an overconcentration on the production of luxury goods. As for other countries, the Belgian provinces were under foreign rule and Switzerland lacked raw materials and was remote from markets. They all cast envious eyes on Britain, a land that was becoming very rich and powerful, and began to make strenuous efforts to acquire her expertise.

Dupin was far from being the first visitor to be bent on learning Britain's secrets; there was a long tradition of people making such journeys behind him. Industrial espionage was a time-honoured means of technology transfer among nations. In the 1700s and 1800s, Britain was the target of numerous visits, both official and clandestine, that were undertaken with the intent of obtaining the secrets of her technological prowess in every field from button-making to steam engines. Given Britain's acknowledged maritime superiority, her dockyards and arsenals were naturally targets for such espionage. Visitors had come to observe Britain's success, both industrial and military.

In the seventeenth century, Peter the Great of Russia had had early contact with English sailors at Archangel, where he took lessons from English and Dutch skippers who visited that port (Pares 1962, 225). In 1689, he visited Holland, but being told that the theory of shipbuilding was better understood in England, he moved on to London. His aim was to learn the art of shipbuilding, working with his own hands in Deptford (226). He was given quarters in Deptford, but it seems that he and his companions left them in a terrible state as a result of their stormy revels.[2]

Many Frenchmen came to Britain, and their visits were not always clandestine. As early as 1764, the French government had sent a young man from Lyon, Gabriel Jars, to study English methods of iron production (Jars 1774–1781). A few years later, the minister for the navy sent Babaud de la Chaussade, the master of the Guérigny forges in the Nivernais, to study industrial progress in several foreign countries, including England (Bourgoign 1971, 6). Bosc d'Antic, a specialist in glass production, came to England from the French Académie des sciences to study British methods. In 1771, he reported to the académie on his visit, calling for support for the French glass industry so that it could compete with Britain.

In 1775, Marchant de la Houlière visited England to investigate the superior quality of English coal and the nature of English iron ores. One of the reasons for the French failure in steel production was their reluctance to use Swedish iron for conversion, as the English did. This would change, however, as Ballot observed: 'The French government, during the second half of the eighteenth century pursued with the most remarkable efforts the importation of this industry into France' (1923, 8).

In 1779, less than five years after the completion of James Watt's first successful steam engine in Britain, the Périer brothers had established a similar engine in France (Blum, Cameron, and Barnes 1966, 67). The construction of a blast furnace designed for coke, the first on the Continent, began at le Creusot in 1782. By 1790, there were some nine hundred spinning jennies in France.

Of course, some visitors to Great Britain came simply out of interest and curiosity, but in the main, their goal was to find information that would help their home country to rival Britain in industry and production. They came on behalf of their own companies as skilled craftsmen, aiming to expand their trade and to acquaint themselves with technological innovations. This was very much the case with the Swedes, Germans, Russians, and after independence, the Americans (see Flinn 1926–1927, 95–109; Robinson 1975; and Jeremy 1981, chs. 4–8). France was particularly interested, and with her wider range of industries, she had much more to gain and fewer scruples with regard to her ancient enemy. Where machine processes were concerned, the French found English workers to be, to quote an official report,

> haughty, quarrelsome risk-takers and greedy. There is nothing easier than to suborn them, and when a new machine produces some gain for English industry the French government can always be master of it in six months for a small outlay (Dupont 1786)

Much has been written about industrial spies, particularly by Harris (1998), but the visits of some of the engineers I shall mention here seem to have gone largely unrecorded.

France was deeply involved in espionage, but the gathering of intelligence about British technology was not always clandestine. It was a European tradition, as Harris observed,

> to receive foreign scientists or savants, members of this academy, or corresponding fellows of that, and give them the opportunity to visit some celebrated mine or manufacture. Foreign army officers were expected to be honourable and gallant fellows of no particular technological skill, but the Baron, the Chevalier, the geologist, the naturalist, the chemist, the aeronaut, the Brigadier, the Captain, was very likely an iron master, an artillery technologist, a member of the bureau of commerce, an inspector of industry, an emissary of a foreign office, or indeed more than one of those things at the same time. (Harris 1985, 127)

Most of the technical innovations associated with Britain's industrial revolution were made by men without formal training. France saw her role in adapting improved processes and techniques and in providing systematic training for engineers. It has been seen that the French system of technical schools was already well established when the Napoleonic period began. It was from one of these schools, the Ecole des ponts et chaussées, that observers began to arrive in Britain in the 1780s (see Bradley 1992 and 2005). Reports of these visits were collected and preserved in the library of that prestigious school. Were some of these engineers spying for France, and if so, who was behind it? It must, of course, be stressed that the main purpose for most of them was study and observation, but the fact remains that their reports could be very useful to an enemy of England. So, a new type of Frenchman was coming to the shores of England; they were not industrialists bent on gaining the secrets of Britain's success and using them to their own ends but ostensibly, at least—servants of the state on educational visits. Their reports provide details of what impressed them, and their comments also give clues as to the usefulness to the French state of their accounts. Despite the temporary and uneasy state of peace, what could be potentially more useful to France than detailed drawings of bridges, ports, arsenals, and the road network?

The same applied to the official visits of representatives from, for example, the French navy. Forfait and Lescallier, who came to England in 1798, produced a very lengthy report of all that they had seen (Bradley 2005). They were not the first or the only visitors from the French military, but coming as they did just prior to the Revolution, closely followed by Napoleonic imperialism, the details they recorded must have proved to be very pertinent. Their journey was similar to those on which Charles Dupin would embark in the next century.

A significant effect of the Napoleonic period of military adventure was to cut off not only France but all of continental Europe from active intercourse with Britain. To be sure, as Henderson (1966) showed, the French continued to procure new machines even in time of war, yet there can be little doubt that the transfer of technological knowledge across the Channel was severely hampered until after 1815. However, some visits continued. In 1814, the economist Jean-Baptiste Say (1767–1832), a future colleague of Dupin's at the Conservatoire des arts et métiers as a professor of industrial economics, was able to describe the immense changes that had taken place since the period prior to the French Revolution. He wrote,

> Everywhere the number of steam engines has multiplied prodigiously. Thirty years ago, there were only two or three of them in London; now there are thousands. There are hundreds of them in the large manufacturing towns and they are even to be seen in the countryside. Industrial activity can no longer be profitably sustained without the powerful aid they provide. (Say 1815, 30–31)

More will be said of Say when I look at Dupin's work in economics.

After Waterloo, Frenchmen joined the growing number of visitors to Britain, often as tourists in pursuit of the wild scenery, but many came in the interests of French industry. In 1817, Georges Dufaud, a *polytechnicien* and the director of the forges of Trézy and Grossouvre in the Cher, was sent to buy British iron 'for use in France to ensure the production of a quality equal to that of the British iron and steel industry' (Le Mallier 1971, 31). British technical knowledge and experience were also of great significance in the early development of French railways; Louis de Gallois, also a *polytechnicien*, stayed for sixteen months between 1817 and 1818, studying British coal mines, foundries, and forges as well as railways. On his return to France, he presented a paper to the Conseil Général des Mines, giving a detailed description of railways in British mines and recommending their use in France (Gallois 1818, 137–144).

7.2. Dupin's Visit

Charles Dupin's visit was in advance of those made by these two men. He intended to make the most of his journey and to see as much as possible. He immediately applied himself to studying the English language and local customs. If he did not yet know England, he certainly did not know the language, which he had never spoken, but he would use a dictionary and learn it gradually by direct contact, spurred on as he was by his desire to increase his maritime knowledge and to complete his *Tableau de l'architecture navale* (Survey of naval architecture). This first visit, which lasted six months, would be concerned only with England; during the course of his second visit, he would travel the whole of Great Britain.

To his elder brother, Dupin described his first experiences, writing from London on 2 August 1816,

> I have arrived in London in good health. I have found an inexpensive room and am living on my grant; this way, I'll cope.
>
> Everywhere I look, I see wonderful things in a town four leagues long and ten wide. The people who, we were told, hated us, are extremely amiable and never insult us. Here is my address:

> Mr Dupin, at M. Edden, taylor [*sic*] and draper, Christchurch passage, Newgate Street, London. (DRB; PJ 43)

At the foot of the letter, he had drawn a gibbet with a condemned man on the end of a rope.

Among Dupin's close friends were the Pronys. Prony was the director of the Ecole des ponts et chaussées, and his wife was a popular, cultured woman and a talented musician (see Bradley 1998). Prony was probably France's leading engineering educator during the first half of the nineteenth century, a man well respected internationally. He played a special part in the development of scientific and mathematical education during the period from 1790 to 1840 and was the principal teacher and examiner at the Ecole polytechnique from its inception. A very influential figure in the Académie des sciences and a good friend to many budding scientists, his was an acquaintance well worth cultivating. It was to the Pronys that Dupin wrote, also on 2 August 1816, to describe his new lodgings, and the tone of the letter indicates the warmth of their friendship:

> As all the joys of this world are brief and ephemeral, I am about to leave the prettiest home in the world; to my right I had Newgate prison, opposite, a church, and beside me, a cemetery planted with trees in zigzag and interspersed with tombstones! Finally, on the left a hospital. There you have, Madame, a dwelling worthy of the somber Young and gloomy Harvey.[3] (BN Naf 3282; PJ 44)

Dupin continued with his description of 'the island of melancholy' and the ever-present clouds of coal dust, and then turned to the subject of English women: 'Mesdames, the English ladies, have a pretty face, fine bosom, big feet, large back and a long neck'.

He was about to leave for the provinces, and he provided details about his wardrobe, one of his constant preoccupations:

> I wanted to buy green breeches and gaiters, with a long coat to my feet, so I wouldn't look like a foreigner but, reflecting that the way I speak and above all, understand what people say to me, that is to say, nothing, would destroy the effect of my outfit, I gave up on the idea.

Asking forgiveness for his trivia, he added, 'That is why I have very little inclination to write novels, although my life is one, or rather, contains half a dozen' (BN Naf 3282; PJ 44).

On 1 September, Dupin wrote from London to the ministre de la marine,

> The French Ambassador, to whom you had recommended me, has taken much interest in the success of my requests and he has been twice himself to the British Admiralty; he has written in my favour. At present, Sir, you may attribute to my love for the arts and sciences the care I shall take to make my journey fruitful. (ASHM CC7 779; PJ 45)

On the same day, to a person unspecified (perhaps Mathieu), he wrote of the success of his overtures, 'I have succeeded in everything I wanted. I have permission to see all that I wish to observe. I think that no traveller has ever been treated in such a favourable way' (ADN; PJ 46).

Then, on 2 October 1816 from Portsmouth, Dupin recounted some details to his elder brother, complaining, as usual, about his shortage of money:

> I imagine that you will read my detailed itinerary in my letter to my father. You'll see I'm not being idle and that, if I return without a great deal of knowledge, it will not be because I haven't worked hard.
>
> I have to travel cheaply so as not to eat in three months what has to last me for six. So I have sat on the top of English stage-coaches. I took a seat in a respectable and advantageous corner, on the right at the front, to see the country and to have my right hand free. I travel with suit, frock-coat, cape and umbrella, my dictionary in one pocket and the alphabetic description of the towns and counties in the other. (ADN; PJ 47)

On 5 November, he again wrote to his elder brother,

> It is now almost three months since I had news of you and you know what pleasure I shall have in hearing from you. We are six

> postal deliveries behind because of bad weather. Perhaps something will come from you. I still need to have 800 of the 1,200 that you delivered to M. Lafitte. You should tell my father that if ever he decides to give me the great present he's been talking about for so long, it is now that 5 or 600 F would ease my path. (ADN; PJ 48)

7.3. The Return

Dupin returned to France on 22 January 1817, six months after his departure. He had made a mass of notes which, together with those of his next visit, would enable him to complete his *Tableau de l'architecture navale* and to draft various other well-researched works. The day after his return, he submitted to Jean-Pierre Edmond Jurien de la Gravière, conseiller d'Etat and director of ports and arsenals, a report of his journey which would be presented to the Académie des sciences on 26 January 1818 and reproduced at the beginning of his *Mémoires sur la marine et les ponts-et-chaussées de France et d'Angleterre*. He provided a list of the places he had visited during his stay:

> London, docks of the West India Company and the East India Company.
>
> Southwark.
>
> Liverpool: docks, rope making.
>
> Bristol: the dock.
>
> Plymouth: three visits to the naval arsenal, one visit to the Victualling Office (Food supply service for ships) and one to a naval hospital.
>
> Portsmouth: Four visits to the naval arsenal, visits to ships and the young naval officers' school.
>
> At the mouth of the Thames, a visit to Chatham, Sheerness, Isle of Sheppey, *Bellerophon* ['Billy Ruffian', as the British sailors called her (see Cordingly 2004)], which had taken Napoleon to Saint Helena the year before, now transformed into a floating prison for convicts.
>
> Woolwich: Visit to the naval artillery arsenals.
>
> Greenwich: The Observatory and the naval hospital
>
> Deptford: The Arsenal.

In a manuscript of this report, dedicated to Sganzin (J. M. Sganzin [1750–1837], a *Ponts et chaussées* [bridges and highways] engineer who worked for a long time with Prony), Dupin explained that he had studied the mechanical and theoretical procedures of the British navy as well as its administration. In England, he said, he had been able to see the methods to establish harmony between the most important branch of public power (the navy) and the form and workings of a representative and monarchist government (Dupin 1818b). Dupin was a constant admirer of England's more liberal constitution. He added that he had been permitted to visit the military arsenals in Plymouth, Portsmouth, and Sheerness and subsequently, those of Deptford and Woolwich. According to this report, he had managed to receive all the information he wanted and nothing had been hidden from him, not even future plans. It had been possible for him to study the commercial system, see the dockyards, and even attend the launching of a ship—the ceremony for which, Dupin observed, was similar to that for a warship.

With his social and economic interests, he visited hospitals and prisons in Liverpool and Bristol, the canal which crosses the full length of the city, naval dockyards, and the galleys heated by water vapour that were so advantageous to the navy. He had also visited the factory in Birmingham where these galleys were produced (Dupin 1818b). In Plymouth, he paid three visits to the naval arsenal, where he examined in minute detail all types of vessels as well as the naval hospital. There, the leading surgeon informed him 'on the local advantages and internal discipline'. He further commented that the engineer in charge of work on the Plymouth jetty, built on the pattern of the Cherbourg jetty, had explained to him, on the site, the plan of operations and the means that were employed to overcome the many problems inherent in such work. Dupin arrived in Plymouth shortly after a British engineer at the Chatham naval dockyard had performed an experiment aimed at raising masts with the minimum of effort, and the experiment was repeated for him.

At Portsmouth, Dupin four times visited the naval arsenal, where he saw machines for making pulley blocks. This Portsmouth block-making machinery has been described as 'the first instance of the use of machine tools for mass production' (Gilbert 1965, 1). In the Portsmouth system, operating in 1808, 'relatively few, less-skilled workers were able to turn out large numbers of uniform wooden and metal parts for assembly into low-cost products of standardized design' (Cooper 1984, 225). Dupin observed this process, along with the different types of English construction. Further, he reported that 'a teacher was deputed to show me and explain the local situation, the order and type of teaching in the schools for young naval officers and engineers in Portsmouth'.

The list continues of all that Dupin was able to see in a country that had been so recently at war with France. Chatham opened the doors of the naval arsenal, the barracks, and the practical school for miners and sappers. Dupin was particularly intrigued by Sheerness and the problems of building docks and quays for a completely new arsenal in a place where 'the solid ground is 40 feet below at high tide and more than 20 feet below at low tide'. After describing all he had seen at Sheerness, he commented on some experiments at Woolwich in the field of ballistics and the measurement of the force of cannonballs.

On the Thames, he saw floating dredgers activated by steam pumps (*pompe à feu*), but he was unable to quote the names 'of all the machines and remarkable constructions I examined in each of the establishments I visited'. Hence, he decided to write a complete history of his journey, with a second treatise on all of the civil engineering work he had seen. He ended his report,

> Special circumstances, many powerful recommendations, much patience and perseverance, sharp eyes to see what I was shown, much more luck than merit, that is what made my journey a success. If the results can be useful to my country I shall not regret the constant hardships, nor the occasional vexations I had to endure, nor the sacrifices I had to make in order to achieve my end, that of learning and knowing about the British navy. (Dupin 1818b)

Dupin claimed that he had been well received everywhere—but had he really seen everything? He had no way of knowing. One commentator maintained that the young engineer had great difficulty getting into the arsenals and that he was not permitted to make a single sketch or to take notes; rather, it was his amazing memory that enabled him to retain all that he had so hastily visited (Lacaine and Laurent 1847, 4, 280).

The report of Dupin's first journey to England was presented to the Académie des sciences on 26 January 1818, then reproduced at the beginning of his *Mémoires sur la marine et les ponts et chaussées de France et d'Angleterre* (1818). As for the 'constant hardships', his elder brother was again called to the rescue with a request to 'settle my expenses'. On 23 January 1817, Charles wrote to Dupin *aîné*,

> I left Dunkirk on 22 July 1816 and returned there on 23 January 1817, which makes exactly 6 months. I am asking you to settle my expenses. I shall collect here a small sum due to me with which I shall repay you immediately, whilst thanking you, my good friend, for the advances you have made to me. I am returning, my dear friend, with a rich collection of observations, plans and experiences which three months will hardly be enough to put in order and make presentable … The French ambassador to whom I had been very coldly recommended and who had therefore received me accordingly, ended by taking an interest in my journey when he saw how successful it had been. He received me at his home and often asked me to dine with him. On my visit before last, he asked me if Marshal Ney's[4] lawyer was related to me:
>
> Brother, *M. l'ambassadeur*.
>
> He is highly talented.
>
> You are too kind.
>
> But I fear his principles are too republican.
>
> On the contrary, Monsieur, and I have frequently seen him vehemently rail against that system; what is more, he was most unpopular with Bonaparte's provisional government.
>
> Ah, you give me pleasure and I am charmed by what you say.

> A moment later, a certain Lady Clainger, with whom I had dined at the ambassador's residence, asked me aloud if I were his brother.
>
> Yes, Madam.
>
> Ah then, Monsieur, I am a great enthusiast of your brother's; I was present at the defence of the three Englishmen which gives him infinite distinction and I would have embraced him had I been able to do so. Please tell him that on my behalf.
>
> I had written to you to ask you to send by the stagecoach, 40 copies of my mathematics papers of which M. Gergonne, of Nîmes or of Montpellier, wrote that he had sent you a hundred copies. (DRB; PJ 49)

The tale of the three Englishmen Dupin mentioned is famous in French history. The Count de Lavalette, the administrator of the post service under Napoleon, was arrested for conspiring with him and condemned to death on 21 November 1815. On 20 December, Lavalette's wife and daughter visited him in the Conciergerie, and he managed to dupe the guards by escaping in his wife's clothing, leaving her in his cell. He left Paris on 8 January 1816 with the assistance of three British officers, Michael Bruce, Captain Hely Hutchinson, 3rd Earl of Donoughmore, and General Robert Wilson—the three men who would be defended by Dupin *aîné*. In 1822, Lavalette was pardoned and returned to Paris, where he rejoined his wife, who had gone mad as a result of her detention (see 'Antoine-Marie-Chamans de Lavalette' in Mullié 1852). For their part in the adventure, Wilson and Hutchinson were sentenced to three months' imprisonment.

Dupin also mentioned J. D. Gergonne (1771–1859). He was a French geometer of note who in 1810 had founded the Annales de mathématiques pures et appliquées, subsequently the Annales de Gergonne. He specialized in projective geometry, and in 1816, he published a solution to the problem of Apollonius. In the same year, he was appointed to the chair of astronomy at Montpellier. Many great mathematicians published in his journal.

7.4. 'LA STRUCTURE DES VAISSEAUX ANGLAIS'

Dupin's stay in England was very important for his publications. He began his 'Structure des vaisseaux anglais considérée dans ses derniers perfectionnements (The structure of English ships with their most recent improvements) by pointing out that he had been attracted to England by the desire 'to make more complete and less imperfect' his *Tableau de l'architecture navale aux 18e et 19e siècles* and that he had been warmly welcomed everywhere. He then stated that it was Laplace who had advised him to carry out a reasoned investigation of the progress achieved in Great Britain in the naval field. He spoke of Robert Seppings (see Lambert 2006), who had introduced in England a means of providing the timber of ships with new strength; these principles, applied successfully to the refurbishing of various English vessels (*Tremendous*, *Ramillies*, and *Albion*), had received praise from the most distinguished specialists (Sir Joseph Banks, Dr. William Hyde Wollaston, and so on). However, Dupin added that French builders in history had known so well the truth of the principle reproduced by Seppings that they had used it precisely to achieve the same result for the strengthening of ships and preventing flexing.

This paper by Dupin appeared in French in the 1817 proceedings of the Royal Society under its exact title, 'De la structure des vaisseaux anglais, considérée dans ses derniers perfectionnements' (On the structure of English ships with their most recent improvements). It was a rare honour to have a foreign paper published in this journal, and in its original language.

Dupin summarized his work:

Contents of the paper:

a. An historical survey of French and English research on the improvement of the strength of ships.
b. Innovations introduced by Mr Seppings, British engineer and ship-builder at Chatham, as follows:
   The filling of all the gaps between two adjacent ribs below the orlop:

Suppression of the inner planking:
Replacement of the reinforcing ribs by oblique ribs:
c. General considérations and a study with supporting calculations, of the last two of the five points considered.

The paper had previously been the subject of a communication by Sir Joseph Banks to the Royal Society on 19 September 1816, a month before Dupin's return to Dunkirk, and it was to be the basis of the fifth and last paper in his *Applications de géométrie*, published in 1822. The 'Structure' contains references to results obtained in Dupin's 1813 study of wood strengths ('La flexibilité et la force des bois') as well as to the beginning of the 1814 *Tableau de l'architecture navale*. As indicated in the table of contents, the paper includes a summary of the French and English studies on the improvement of the strength of oceangoing vessels and on the innovations of Seppings, a British engineer at the Chatham naval dockyard and author of 'On the Great Strength Given to Ships of War by the Application of Diagonal Braces' (1818).

That innovation by Seppings had been reported to the Royal Society in 1814 (Seppings 1814; Lambert 1991, 61); it was based on the 5 bar gate principle, which Seppings found to be the strongest structure for ships. He had observed that a five-bar gate was strengthened by fixing a diagonal strip of wood across it; a similar process applied to ships would enhance their strength and durability. Reporting his findings to the Royal Society, Seppings illustrated this:

> The length of a 74 gun ship being 170 feet or more, it requires little knowledge of the strength of timber to perceive that planking of the length, however thick, or in whatever way joined or put together, must, under the present system, bend with its own weight. The fastenings, and consequently the connection of the several parts of the fabric, must therefore suffer for want of *stiffness*, and a change of form is the consequence.
>
> This may be shewn by putting together four pieces of wood, and securing them with iron pins in the form of a square, which on the least pressure may be made to change its form on the rhombus; but let another piece be fixed to it diagonally, and the figure of the

> frame will be found immoveable. Place a bar in the middle parallel to two of the sides, and secure it firmly by iron pins, still the figure will be easily moved by hand, like a parallel ruler … but apply to the frame what the carpenters term the brace in a common field gate … and the figure remains, as before, immoveable. The greater the length of the frames, the greater will be the advantages of the new principle, both in stiffness and strength. (1814, 289–290)

Dupin claimed that several of the principles expounded by Seppings had been known for a long time in France; in evidence of this, he quoted the work of Henri-Louis Duhamel de Monceau (1708–1782), an inspecteur-général in the French navy. He also referred to Daniel Chauchot, a naval engineer, and to another engineer, Antoine Groignard, who had experimented with the use of ceiling plates and inner planks. He ended with the remark that Seppings's principles were new neither in practice nor in theory, although he acknowledged that he had been responsible for certain innovations, such as the use of the web frame. According to Fox,

> Dupin's comments on British naval establishments provoked the translator of his Mémoires sur la Marine into adding a tart introduction and sustaining a critical running commentary in a series of footnotes. In these additions, Dupin was accused not only of factual errors and unfounded assertions but also of an arrogant pride in French scientific superiority which ignored and, by implication, disparaged the practical skill and inventiveness of British shipbuilders and engineers.[5] (Fox 1985, 315–316)

Dupin's ability to absorb and remember details is evident from his 'Notice sur les travaux de breakwater ou de jettee [*sic*] en construction dans la rade de Plymouth, 1816' (Work on a breakwater or jetty under construction in Plymouth Sound; ENPC ms 1877). In this manuscript, he explained that the Plymouth structures were particularly useful because of the general problems of the terrain. The Plym and the Tamar form two vast basins at their mouths, and their waters meet in an even larger basin, Plymouth Sound. Commercial establishments and the town of Plymouth were situated on the right bank of the Plym, with the naval establishments and the Plymouth dockyard on the left bank of the Tamar. The two towns were linked by another, Stonehouse, and the entire agglomeration contained some sixty thousand inhabitants. Dupin provided a complete picture of the area, then proceeded with an account of the problems involved in building the breakwater. He concluded by noting that

> the works I have just described are directed by Mr. Whidbey [*sic*],[6] a former navigator who accompanied Vancouver round the world, who is a student of the famous engineer Rennie,[7] the person responsible for the plan and general direction of operations.

Apparently it was John Rennie himself who recommended that Whitby should show Dupin around the works and explain everything to him.

Dupin sent his report to Prony, the director of the Ecole des ponts et chaussées, expressing the wish that Prony should submit it to the Académie des sciences, preferably when Rennie was present. He concluded,

> Without your recommendation and that of Mr Rennie to Mr Whidbey [*sic*], I could never have given such an extensive and, may I say, satisfactory idea of a work which pays tribute to Mr Rennie ... I am continuing my journey, full of admiration for a host of beautiful things which strike my gaze everywhere. (ENPC, ms 1877; PJ 50)

With all of the research carried out by Dupin and others, it is strange that no request for a patent has been found either in his name or that of the state. In fact, there were very few applications for patents at all in the naval field in France. They would have been deposited in the archives of the Institut national de la propriété industrielle in Paris; one in particular is to be found there. This was from an English architect and inventor, living at Fenchurch Buildings in the city of London, John Walters, requesting a ten-year patent for an invention he had made in England for 'improvements in the construction as well as the refurbishment of ships' (Paris, Archives of the Institut national de la propriété industrielle). (For an extract from Walters's application and details of his invention, see PJ 51.) From his application for a patent, it is apparent that he was working on diagonal strapping, in much the same area as Seppings. Originally patented in England on 7 November 1814, Walters's work received a French patent on 25 April 1815. The request must have been submitted during the first Restoration and during Napoleon's exile on Elba.

Endnotes

1. Much of this chapter has been published in Bradley and Perrin 1991 and Bradley 2008.
2. Peter's unfortunate hosts were the family of the diarist John Evelyn, at Sayes Court.
3. The 'somber [Thomas] Young' and Hervey were British scientists. Young would demonstrate in 1817 that light waves must be transversal vibrations. Auguste-Jean Fresnel (1788–1827) had been working in Paris in the same field.
4. Marshal Ney, a soldier of the Revolution and the Empire, was accused and condemned in 1815 for treason against the king and the state.
5. Dupin's remarks were not entirely disparaging. He acknowledged that he and Seppings had achieved similar results in their calculations.
6. Thomas Whitby, a famous navigator.
7. John Rennie, a great Scottish engineer who was responsible for the construction of Waterloo Bridge, London Bridge, factories, ports, and canals.

Chapter 8

Meanwhile, in Dunkirk

8.1. Dupin's Return to France

Charles Dupin went home to France after six months of intensive activity in England. He had now to complete the account of all his meetings and visits. He had acquired a good general knowledge of the British navy, but to probe further, he was already contemplating making another visit as soon as possible. The Académie des sciences would be welcoming him in before long (in September 1818), and for his election to that illustrious body, Dupin would know how to make the most of his missions to Britain for his nomination. He was to leave again for that country at the end of May, just four months after his return from his first mission. In the intervening months, he would pursue his engineering duties in Dunkirk whilst awaiting permission to leave and paving the way for the second mission he was so eager to undertake, as well as for his entry to the Académie des sciences. His 1813 *Développements de géométrie* would serve him well in the latter project, and his election would prove to be an excellent qualification for his appointment as a professor at the Conservatoire des arts et métiers in 1819.

French investment in the navy had been far outstripped by the vast sums allocated by the British to their service; French and British expenditure was at a 1 to 3 ratio in 1800. This changed to a 1 to 2 ratio in 1801 and 1802. However, the gap increased again during the ensuing years, reaching 1 to 4 in 1806, when the British naval budget was attaining its highest level. The British naval supremacy was established (see Bradley 1994 and Villiers 2005). According to Villiers, Napoleon 'never fundamentally understood what a navy equipped for war could do, nor its limitations (Villiers 2005, 7). It is true that he created *préfets maritimes* (officials responsible for the navy and shipbuilding), but it was not until the end of the nineteenth century that any real governmental attention was given to the French navy.

During the Restoration, purges ensued in the Corps de la marine under the new ministry of François Joseph de Gratet, Vicomte Dubouchage. Dupin managed to escape the effects as a result of the influence of his friend Tupinier (see chapter 4.3), who had directed the ports and arsenals since 1814. In 1816, Tupinier, suspected of Bonapartism, was sent to the forestry service in Angoulême. There, apparently, his correspondence was scrutinized by an ultraroyalist committee (Lutun 1994, 158). Nothing is to be found in the archives on this episode and his banishment to the forestry service; there is only Tupinier's report and a reference by Dupin in his 'Eloge de M. le Baron Tupinier' (3 December 1850, in *Le Moniteur*).

In this climate, Dupin probably considered it politic to accept his appointment to Dunkirk without demur, but when he arrived there, he found the port in a state of neglect with 'not a nail or piece of timber' (Dupin 1819b). As Junges (2009) put it, both engineers must have considered their appointments to be a bad joke. Alone responsible for the Génie maritime service, Dupin was to supervise the reequipping of various ships and the changing of Dunkirk to a commercial port (Dupin 1817).

He continued to complain about his finances, for instance in a letter to his long-suffering elder brother on 3 February 1817:

> I have here two orders for 660 F. I shall keep 60 which, with the other 50 in my pocket, will provide enough for me to eat and drink until further notice. I shall think of a way to send you your 600 F. You will receive it right away … You'll see that when you have received the 600 F I'm going to send you, my debts will be more than paid off. The rest will be to settle a small sum I owe Mme Courcier.[1] (DRB; PJ 52)

This letter was followed by another, again to his elder brother, dated 9 February 1817, in which Dupin referred to the report of his first visit to England and to his engineering activities in Dunkirk:

> I have received a very pleasant letter from Admiral Rosily, to whom I had sent the description of work to produce large maps of England. As Admiral Rosily heads the department of naval maps and plans, he could only be pleased to know of what the English are doing in a similar area to his … I am completing the first part of the account of my journey. This paper,[2] which includes everything related to the artillery will be the most likely to interest the minister as general of the artillery … I have just received the maps and books I bought in England to provide material for my work. The Dunkirk Customs, in favour of the 'science of the a + b', chose not to charge me the duty, which would have completed my ruination.
>
> Next Sunday, I launch my corvette with all the solemnity necessary at such an event. I did not want to end this letter until today, so I could tell you that my corvette was launched with great success. I enclose a note on this truly imposing ceremony and you can insert it in the *Moniteur*.[3]
>
> M. Sané [Jacques-Noel Sané, the highly respected naval architect; see chapter 5.2] has just written to me that the minister, recognizing the usefulness and importance of the results of my journey, has given me a *Génie maritime* student whom he will make cover 300 leagues to help me design the plans which will accompany

> my reports … I shall enjoy a little peace and quiet now that my corvette is launched and that will be to the benefit of the account of all that I saw in England. (DRB; PJ 53)

Dupin constantly returned to money matters in his correspondence. In another letter to Dupin *aîné* on 27 February 1817, he wrote,

> Of the money you have, I must pay almost 300 F to Mme Courcier, as soon as I have settled the cost of the books she sent me in England. This sum of 800 F I took with me originally, plus my journey from Paris to Dunkirk, plus 2,400, constitute the real expenses of my journey, without counting the deterioration of my wardrobe. (DRB; PJ 54)

From this letter, it is apparent that for his first trip to England, Dupin spent more than 3,200 francs, without counting what he spent on the journey from Paris to Dunkirk. In the same year, Georges Dufaud, an ironmaster who spent three months in England with invitations of a similar kind, spent only about 600 francs (Le Mallier 1971, 31). Dupin's allowance from the Dunkirk administration of which he complained so bitterly was therefore not so miserly after all, and he had spent much more than Dufaud.

Returning to the letter to his eldest brother of 27 February 1817, Dupin continued with the details of his work. He explained that Sané had told him to return to Toulon as soon as his corvette was launched. However, he had first to submit his accounts for the years 1816 and 1817, complete the editing of his account of his journey to England, return for three months to England (provided that he was paid for his first trip), and publish the account of his journey. He then added,

> The Royal Society of London has just greatly honoured me; it is publishing in its 'Philosophical Transactions' the paper I wrote on shipbuilding ['La construction des vaisseaux'] before coming here. I think that, since Franklin, who was half-English, no foreigner was thus honoured. By that, I shall prove to the English that I know how to appreciate their liberal welcome and, to the French, that my work is appreciated abroad. (DRB; PJ 54)

As evidence of his gratitude to his English hosts, Dupin again wrote to his elder brother from Dunkirk on 2 March 1817,

> Please entertain Dr Young, secretary of the Royal Society of London and one of the most distinguished scholars of the British capital. He is coming to Paris … He is one of the savants whom I found the most obliging and he gave me essential help. (DRB; PJ 52)

Meanwhile, Dupin was finding irksome the constant transfer from one place to another in the course of his duties. On 22 March 1817, he wrote to Dupin *aîné*,

> Sané wants me to be already in Toulon. When I am in Paris, he is eager to exile me in the provinces and, when he has pushed me off to the north, he pushes me back to the south. Try to moderate this 'transfer fever' in M. Sané. Since the minister has sent an officer to help me, it is obvious that he wanted me to complete the report of my journey, which will make an enormous in 4° [volume in quarto], of which I have already given him the first third; the second third will be ready in ten days and the last in two months, at the moment when I would wish to return to London to finish many important things and write a work which would establish my reputation. (DRB; PJ 56)

It would appear that Dupin felt able to ask anything of his brother.

8.2. Dupin Pursues Election to the Académie des Sciences

Dupin was determined to fulfil his desire for election to the académie, as illustrated by the following letters. In a letter to the physicist Jacques Charles[4] from Dunkirk on 15 April 1817, he wrote,

> Emboldened by the manner in which the *Institut* received my work, particularly under your presidency, I take the liberty of putting myself forward to replace M. Rochon[5] in the physics section.

> I attach to this letter an account of my work; may it deserve your honourable approbation. I respect you too much to add anything about what my work may say in my favour; I must rely on it with complete confidence in your impartiality.

P.S. May I presume to ask you to forward the enclosed letter to those of your colleagues to whom they are addressed? (ADN; PJ 57)

Dupin was to be elected to the mechanics section to replace Périer, the industrialist and inventor, following a further letter of application of 6 September 1818. He continued to enlist the help of his elder brother in his plans, and on 15 April 1817, again from Dunkirk, he wrote,

> I am asking to be put on the list of candidates from whom will be chosen at the *Institut* the successor to the physicist Rochon who has just died. I send herewith a letter from M Lefebvre-Joneau with an account of my work which I ask you to read. If you could go and see M. Lefebvre, it would be very important to me: he is one of the five members of the section which must draw up and present the list of candidates. The others are Poisson,[6] Gay-Lussac[7] and Charles. Recommend me to M. Dayeux (show him the account I'm sending you to make the best possible use of it) and to other members of the *Académie des sciences*, if you know any others. You know that the future of my life depends on my election to the *Institut*: the whole complexion of my affairs would then change.
>
> However, were I not elected, it would still extremely important to me to be presented and to have a certain number of votes: that would prove that at the first presentation, I shall be chosen forthwith.
>
> P.S. I have sent to Maréchal Marmont,[8] *académicien libre*, the analytical table of my paper on the English artillery. He replied, 'The work you bring to my attention must be of great importance; if you would be so good as to send it to me, I shall gladly present it to the *Académie*'. (DRB; PJ 58)

Dupin was leaving no stone unturned. On 2 May 1817, he again badgered his elder brother. This time it was to deliver Marshal Auguste Marmont his paper on the English artillery, which constituted a third of his account of his journey to England. He explained that he had already sent it to the minister for the navy along with the second part on hydraulic architecture and that he would soon have completed his *architecture navale* (the work described in chapter 10.1). He then asked his brother to look at his *description de l'école de Woolich* [*sic*] (description of the Woolwich school), the highly successful institution, which he would see again on his next visit. He continued,

> The minister has decided that I shall return to England. I impatiently await his permission. I should like to leave between the 25 and 30 of this month: it is important that I arrive in time so as not to find, as happened last year, all the English disappeared and away from London. M. Sané seems to be well disposed towards me: I have received from M. Rolland, inspecteur général adjoint, a very flattering letter concerning my work on the artillery.
>
> P.S. As an abundance of good things can do no harm, if you find another good recommendation for London, send it to me. (DRB; PJ 59)

Dupin also asked a lot of his good friend Bajot. On 19 May, he wrote to him with

> a review of the trigonometric measure[9] concerning the coasts of Britain, for insertion in your annals. When I am in London, I think it would be good idea for you to send me two or three copies of the edition with that article. I shall then use it to interest directly in your work the governor of the Woolwich school, who seemed to be much flattered by this review which is of particular interest to him. You know that personal interest is the greatest incentive for poor humanity and men of letters greatly appreciate journals which speak well of them. They would like to see them universally distributed. (ADN; PJ 60U)

Dupin added that he had just received a letter from Bajot with notice of his departure date for England, and asked him, Bajot, to remain his good friend. He, indeed, ended his letter, 'Je vous embrasse de tout mon coeur' (I embrace you with all my heart)—an indication of the new strength of their friendship. Dupin added a postscript with a request to forward to him a packet from Mme Courcier via the French consul general in London. He explained that 'it concerns matters important to the success of my research and you take such a friendly interest in it that I need not fear that I exhaust your kindness' (ADN; PJ 60).

The Royal Military Academy at Woolwich had had a particular effect on Dupin. Founded in 1741 to train artillery and engineer officers, most of the technical development was carried out by serving officers and civilian employees. The written results were rarely published, but they were kept in manuscript form—a wise precaution against espionage. Similarly, for the students, there were very few manuals or texts; the onus was on the young officer cadets to make notes and compile them. According to Graves (2006), entrants in 1793 were to study fortification, artillery, mathematics, drawing, the French language, and chemistry, as stipulated by the academy syllabus for that year. The gunners were required to spend at least two hours a day on quick-firing drill throughout the period of the American War of Independence and the Napoleonic wars. No official manuals of gun drill were published until the middle of the nineteenth century. These methods must have been of great interest to Dupin the educator, and it is small wonder that he hoped to return to Woolwich. Moreover, the secrecy of the academy was, as Graves has said,

> in distinct contrast to the general practice in Europe where innovations in artillery often found their way into print a few years after they were formulated. European, particularly French, artillery theorists appear to have been unaware of the manuscript holdings at the RMA but, unable to consult it, were reduced to complaining about 'the mysteries of the school at Woolwich'. (Graves 2006; see also Tousard 1813)

8.3. Again to Britain

On 19 May, Dupin heard from Bajot that he was going to leave once more for Britain, and five days later, he announced triumphantly to his brother, 'I leave next Monday for London'. He added his inevitable comments on finances:

> I am leaving with the 1200 F they are giving me for the six months of my first visit and one term for which they pay in advance. With some other savings, I shall have enough for my journey. Nevertheless, I thank you with all my heart for the letter of credit you sent me. (DRB; PJ 61)

So, at the end of May 1817, Dupin left again for Britain, having carefully prepared the ground for his election to the Academy of Sciences. His methods and persistence would reap their rewards.

ENDNOTES

1. His publisher (maison Vve Courcier).
2. The reference is to the report of Dupin's first visit, 'Aux moyens de mettre en harmonie la branche la plus importante du pouvoir public avec la forme et l'opération d'un gouvernement représentatif et monarchiste' (The means of harmonizing the most important branch of public power with the form and operation of a representative and constitutional monarchy; ENPC ms 3000), which was reproduced at the beginning of Dupin 1818a.
3. The *Moniteur universel*, or *Gazette nationale*, was launched in 1789 and in 1799 became the official journal for the publication of government acts.
4. Jacques Charles (1746–1823), known for the law on gases which bears his name, dealing with the relationship between the volume of a gas and how it changes with temperature. Charles was closely associated with Montgolfier's research on balloons.
5. Alex-Marie Rochon (1741–1817), known for his work on optical instruments.
6. S. D. Poisson (1781–1840), a mathematician and the author of works on magnetism, mechanics, and optics.
7. J. L. Gay-Lussac (1778–1850), a chemist and physicist, best known for his work on gases.
8. Frédéric Louis Viesse de Marmont (1774–1852). In 1699, with the creation of the académie, Louis XIV had provided for eight *académiciens libres* (free academicians), independent of their specialties. They could be foreigners. In 1785, under Lavoisier's directorship, the number was increased to twelve.
9. This constituted part 6 of the report of his journey, 'Côtes et ports' (Coasts and ports).

Chapter 9

The Second Visit to Britain

9.1. The Departure

Setting off at the end of May 1817 on the second of his six visits to Britain, Dupin began a programme of private visits, going to the Royal Society and taking another look at Deptford and Woolwich (for further details, see Bradley and Perrin 1991 and Fox 1985). This time, his visit would be longer—more than eight months. Moreover, instead of confining himself to England, he was to travel to Scotland, Wales, and even Ireland (although he would see only Dublin). Far more at ease this time, he would benefit from the previous contacts he had made, and he had some useful recommendations from people in high places, although he was as yet unaware of the profound effect this visit would have on the development of his career.

He immediately installed himself in London, visited individuals, and wrote his usual letters mixing important issues with small domestic details. Bajot seems to have become his close correspondent, and one of Dupin's first letters was addressed to 'My dear Bajot'. His correspondence with his friend Mathieu seems to have gone somewhat by the board; perhaps Bajot, in his elevated position, was more useful to him in his plans for career advancement. Dupin wrote on 2 June 1817 that he had visited William Thomas Brande, the secretary of the Royal Society of London and editor of the *Journal of the Sciences and the Arts*, and had suggested the exchange of that journal for Bajot's *Annales maritimes* (ADN; PJ 62), presumably on a regular basis. Dupin had praised Bajot's periodical 'with the same confidence as if I never published in it' and had persuaded Brande to accept the proposed bartering. Dupin proceeded to describe Brande's journal as well conceived and including the works of the London and Edinburgh Royal Societies, the French Institut, and the main European academies as well as analyses of new works. Moreover, the two journals had begun life in the same year, 1816.

Dupin did not miss an opportunity to impress, and he advised his friend to 'send 1816 and 1817 separately to look more impressive and valuable' (ADN; PJ 59). Still in pursuit of the great and influential, he wrote on 13 June 1817 to his elder brother to say that he had sent a copy of his 'Structure des vaisseaux anglais considérée dans ses derniers perfectionnements' to Sané and Laplace; the work had been published in the Royal Society's *Philosophical Transactions*. He had had twenty-five copies made, 'which cost me 80 francs' (DRB; PJ 63)—a small fortune, he lamented.

He was well received by the French ambassador and Armand-Louis-Maurice Séguier[1] (the French consul general in London was a constant support and help to Dupin). He was also delighted with the reception he was accorded by the English people he was meeting again. He had visited two arsenals (Deptford and Woolwich) and intended to leave for Scotland on 15 July (DRB; PJ 63). In fact, he would not leave until the twenty-second, but the reason for this delay is not known.

On 26 June, Dupin wrote to Bajot to the effect that he had been asked to produce an analysis of Bajot's *Annales maritimes* for the '*Journal of the Sciences and the Arts of the Royal Institution* [*sic*] *of London*'. He added,

> I have worked non-stop and, in a first article, I have summarized all your issues from 1816. They will incorporate this article in the issue which will appear within five days and I have, this morning, corrected the proofs they brought me.
>
> I suggest that you put the following article in your issue which is about to appear. The *Journal of the sciences and the arts* has just included a considerably detailed account of the *Annales maritimes* published in 1816, the text of which we shall provide in your next issue. We shall respond to this flattering mark of esteem accorded by England's prime scientific journal, by carefully communicating all of such parts of its contents which may be of interest to the navy. (ADN; PJ 64)

Dupin then added that he would like a note to be included on a different page to the effect that the Royal Society of London, to which he had presented his 'Structure des vaisseaux anglais', 'considered this work to be worthy of inclusion in the collection of its papers'. Even though the article was written in French, it had already been published in the first part of the 1817 *Philosophical Transactions*. Dupin stated, 'We shall provide extracts from this new production of M. Dupin' from this new work, presumably for Bajot's journal at a future date, and he ended his letter in very affectionate terms (ADN; PJ 64).

He was making the most of his achievements and his friends for his own advancement. He was on good terms with the English—always with an eye to his own promotion—and he was doing useful work for the navy. He added a note to Bajot the following day, commenting on the fact that the English were publishing his work in the original French: 'Either they like what I do, or they are just lazy'. He was being well received in high places. On 11 July 1817, he wrote to his elder brother that he had been invited to the home of the Somervilles.

Dr. William Somerville (1771–1860), inspector of military hospitals, fellow of the Royal Society of Edinburgh in 1813, and fellow of the Royal Society of London in 1817, was a great support to his wife Mary in her mathematical studies. A Scotswoman, largely self-taught until her move to London, she made an important contribution to mathematics with her 1832 *Mechanism of the Heavens* and her 1835 *On the Connexion of the Physical Sciences*. As was seen in chapter 3.3, she was among those people who were instrumental in making Laplace's work known in Britain. Although she is not considered to be quite in the league of other women mathematicians such as Maria Agnesi,[2] Marie-Sophie Germain,[3] or Sophie Kovalevskaya,[4] she encouraged Ada Byron, Lady Lovelace,[5] in her mathematical research (Patterson 1983, 148–150) and made her own mark in the mathematical sciences. In his letter to his elder brother, Dupin described Mary Somerville as being 'famous for her vast knowledge of the arts and sciences, including Greek and mathematics' (DRB; PJ 65). He ended by saying that he was about to set off again 'on the route of the wandering Jew', this time to Edinburgh.

On 15 July, he wrote again to tell his brother that the Somervilles were leaving for France, bringing letters for the family. In this letter, there is a rare reference to their younger brother, Philippe. Among the letters for the family, there was one for Philippe. Dupin asked that the Somervilles be accorded hospitality and a welcome (DRB; PJ 66).

He wrote to Bajot on 20 July, still in London, with news of the *Journal of the Sciences and the Arts* and explaining that it would not be possible for him to send further news until he returned to London, when he would write a detailed article on this journal. He was desirous that his 'Mémoire sur les vaisseaux anglais' would be better received than it had been in Paris. He added that he was leaving for Scotland with some forty recommendations for savants, men of letters, engineers, and others. Again, Dupin returned to the subject of money:

> I shall economise my small allowance and as long as one sou is remaining, I shall travel on. I shall return to France only when there is nothing left of the seven thousand francs I set aside for these two journeys. (ADN; PJ 67)

Clearly, Dupin was well connected with the *London Journal of Sciences and the Arts* as well as the *Annales maritimes*, which would serve his interests well.

9.2. To Scotland

On 22 July 1817, Dupin left London for Scotland through Cambridge and via the west coast of England, Hull, Sunderland, Newcastle, and Berwick. By 15 August, he was in Edinburgh, whence, according to a letter he wrote to Bajot, he was about depart to visit the Bell-Rock lighthouse and the Carron foundry. He complained that the weather was dreadful—'a pity, for there are very interesting things to see'. He added that 'if people knew of all the trouble and expense in this land, they would do something for those who sacrifice their money and their health for their country' (ADN; PJ 68).

The next news of Dupin is dated 7 October from Liverpool, in a letter to the minister for the navy, so there is a gap of several weeks. His official report of his visit to Britain was to appear at the end of the year (and will be studied in the ensuing chapters); it would provide more details of his journey but would be, inevitably, less colourful than his personal correspondence. In the 7 October letter, he suggested the navy purchase certain objects which seemed to him to be 'of major importance' (ASHM; PJ 69). These were wrought iron cables, hawsehole fittings, wrought iron hoists for manoeuvring heavy weights, circular saws, and red lenses such as he had seen used in the famous Bell-Rock lighthouse on the Scottish coast. In an undated note from Liverpool to an unnamed correspondent, he said that he had seen a great number of things that would be useful to the navy and that they had been shown to him 'with much less difficulty than in England' (ASHM CC7 779).

Two letters were sent from Dublin on the same day, 22 November 1817, one to Dupin *aîné* and the other to Bajot. Dupin had arrived in Dublin via Holyhead on Anglesey. The report of his second journey to England provided the details of his itinerary, as will be seen. However, he explained to his brother that he had suffered an accident on his arrival in Dublin, and that having at last abandoned his crutches, he was beating a retreat for London and France (DRB; PJ 70). To Bajot), he was much more expansive on his injuries. What had been to his brother a 'sprain' became to Bajot 'broken heels' suffered when, on returning to Dublin at midnight, he had fallen from a height of five feet—this, as he said, after negotiating with impunity the 'mountains and precipices of the Scottish Highlands' (ADN; PJ 71).

Another two letters from Dupin, one to his elder brother and the other to an unknown recipient, refer to his impending return to France. In the first, he complained that he had not heard from his family and added a bitter remark about his father, who had become sous-préfet in Clamecy: 'Do remember me to our father who, absorbed as he is in his love of the *sous-préfecture*, probably no longer thinks of me' (DRB). To the other correspondent, Dupin described his exhaustion from travelling, which had also caused his foot to swell. He was ready to depart as soon as he had made the purchases he was required to make, and he would spend two or three days in Dunkirk before setting out for Paris (ADN).

Dupin returned to Dunkirk on 11 February 1818 after landing in Calais. Prior to his departure, he had written in London in December his 'Rapport général de voyage' (General report of a journey; Dupin 1817). While he was returning to France, the minister for the navy and the colonies received a letter from Séguier, the consul-general in London, dated 8 February and drawing the minister's attention to the quality of Dupin's work during his second visit to Britain. He wrote,

> Charles Dupin has avoided all unnecessary expenditure; after six months, having arrived with no knowledge of the English language, he was able to speak very well, which opened all doors to him. The English are generally uncommunicative and very much reserved on the secrets of their industry. (ASHM CC7 779; PJ 72)

There must be some exaggeration of Dupin's linguistic skills in this. Dupin had previously made a quite lengthy visit to Britain, and he had already demonstrated his ability to learn other languages quickly. As for the opening of all doors, it is apparent that he saw a great deal, but how much was hidden from him cannot be determined. As it was, although the minister wanted him to proceed as soon as possible to Lorient, Dupin asked for four months in Dunkirk to complete his account of the second visit, which he had drafted in the 'Rapport général'. Permission was granted.

9.3. Again in Dunkirk

During the month of his return to France, Dupin published a reply (Dupin 1818c) to a venomous attack on France made by the English Lord Stanhope. As much as he admired and praised Britain and her achievements, Dupin reacted very strongly to Stanhope's speech. However, in his 'Réponse au discours de Milord Stanhope sur l'occupation de la France par l'armée étrangère' [Reply to Lord Stanhope on the occupation of France by the foreign army], he refuted Stanhope's inflammatory statements in measured terms and without animosity.

To understand this exchange, one must recall that three years earlier, after Waterloo, about a million soldiers were sent to occupy the greater part of French soil (fifty-eight départements). The soldiers were British, Russian, Prussian, Dutch, Italian, and Austrian, and they all shared a common hatred of France. They wanted to dismantle the French territory, but the tsar and the Duke of Wellington were opposed to this. Finally, the second Treaty of Paris (November 1815) limited France's territorial losses, but it imposed on the country a huge war compensation debt (700 million francs) and obliged it to support an army of one hundred and fifty thousand men for a period of between three and five years.

At the beginning of 1818, there were but nine months left before the end of the three-year period, and Stanhope found this worrying. He wanted the army of occupation to remain in France to prevent the Jacobins or the Bonapartists from taking power and repeating the excesses of the Revolution and the Empire. Rather than risk such a catastrophe, he wanted to divide France into three states. Dupin replied that Lord Stanhope's fears were unfounded; he claimed that France desired to return neither to the Revolution nor the Empire and that it was satisfied with its government as long as it continued to restore gradually legitimate constitutional freedoms. Dupin felt that the presence of occupying forces would provoke a reaction against England and her allies and a boycott of its trade. He added that he would like to see friendly trading between an independent France and England, both countries being opposed to the prolongation of an unjust treaty. For Dupin, the usefulness of that treaty was a myth inspired by fear, and contrary to the views of Lord Stanhope, the French were not merely awaiting the departure of the army of occupation to overthrow the king.

The 'Réponse' was couched in terms to interest foreign rather than French readers, who were well aware of the tide of events and its effects between 1815 and 1818. Dupin claimed that the French were no longer pursuing boundless freedom and were returning to a wise moderation that was equally removed from the two extremes of the Revolution and the Empire. This pamphlet (Dupin 1818c) appeared in London and Paris, but the French government had the Paris edition seized.

9.4. The Preliminary Report of the Visit

At last Dupin was able to turn to the report that he had prepared before his departure from England. It had been sent by the French Embassy to Count Louis-Mathieu Molé, the minister for the navy, at the beginning of 1818, and presented to the Academy of Sciences in February of that year. It was to be reproduced by Dupin, together with the account of his first journey, in his *Mémoires sur la Marine et les Ponts-et-Chaussées de France et d'Angleterre* (Dupin 1818b), but in much greater detail. It was entitled 'Second voyage dans les ports d'Angleterre, d'Ecosse et d'Irlande, par Charles Dupin, membre correspondant de l'Institut royal de France, Académie des sciences, associé étranger de l'Institut de Naples, des académies royales de Motellier, de Turin, etc., membre de l'Académie ionienne' (A second journey to England, Scotland and Ireland, by Charles Dupin, corresponding member of the Royal French Institute, Academy of sciences, foreign associate of the Institute of Naples, of the royal academies of Montpellier and Turin etc., member of the Ionian academy).

Dupin mentioned Ireland, although he had seen only Dublin, and he did not refer to Wales, which he had visited. Moreover, the Ionian

Academy no longer existed, as Corfu was now a British domain. However, he had made contacts in Ireland, particularly with Lady Morgan (Sydney Owenson), a virulent critic of the Bourbon régime and a staunch proponent of the revolutionary cause.

Although it may seem that Dupin was entirely oriented towards his own advancement and the sciences, he had a love of literature and his country's culture. In his *Lettre à Mylady Morgan sur Racine et Shakespeare* (Letter to Lady Morgan on Racine and Shakespeare; Dupin 1818e), he took issue with the Irish writer and the dismissive attitude towards French classicism in her book *La France*. He fervently defended Jean Racine. This can well be understood. Racine conformed to the strict rules of classical drama and the three unities of time, place, and action, but his portrayal of human passions and foibles has been deeply moving to many generations of readers. Dupin described him as France's greatest poet, who conveyed the human passions with consummate artistry and in great depth (1818e, 5). He told Morgan, 'You have wounded our hearts in what we hold most dear' (8). Far from being totally opposed to the new Romantic movement, when Dupin travelled to Scotland, he visited Walter Scott's home and took delight in that author's descriptions both of nature and of man's achievements. The conflict between Dupin and Lady Morgan ended the friendship they had formed in Dublin, according to Moraud (1954, 128–135).

According to the report of his visit (an eighteen-page manuscript, 'Second voyage dans les ports d'Angleterre, d'Ecosse et d'Irlande'; ASHM CC7 779), the places Dupin had visited were as follows: the Deptford Arsenal (for the second time), Woolwich (again for the second time), London and then the east coast of Britain (leaving London on 22 July 1817), Cambridge, Hull, Sunderland, Newcastle, and Berwick. In Scotland, he saw Edinburgh, Leith, Dundee, Arbroath and the Bell-Rock lighthouse, Montrose, Bervie, Stonehaven, and Aberdeen. He crossed the Highlands via the Moray Firth, Inverness, Fort Augustus, Fort William, along the Caledonian Canal, Loch Lomond, Dumbarton, Glasgow, and Greenock. He returned to England via Carlisle, Liverpool (his second visit), Chester, then Wales and the Ellesmere Aqueduct, Llangollen, Bangor, Holyhead, and thence to Dublin.

Dupin's report consists of eighteen pages and may be seen at the home of the Naval Archives (Château de Vincennes). He provided technical as well as domestic details; he was in no way opposed to relaxation and entertainment. The work makes for good reading. He described, for instance, something that he particularly wanted to see—the launching of a naval warship at the Deptford Arsenal. One of the professors at the Woolwich Artillery School conducted, in Dupin's presence, experiments on the strength of wood—experiments that he claimed confirmed Dupin's own theory propounded in 1815 (see chapter 3.4) and published as *Expériences sur la flexibilité, la force et l'élasticité des bois*. This professor was Peter Barlow, who published his own results supporting Dupin's in 1817 (Barlow 1817). Perhaps Dupin had spoken to Barlow about his experiments on wood during his first visit.

The report then described how Dupin proceeded to Cambridge, where he met the university's vice-chancellor, James Wood, and Robert Woodhouse (1773–1827), one of the few men acquainted with French developments in mathematics (see Ruffieux 2005 and Philips 2000), whom Dupin greatly admired. The Cambridge mathematician Woodhouse's knowledge of Continental mathematics was described by Grattan-Guinness:

> [He] showed good knowledge of it in his history … of the calculus of variations … He may have written (some of) the reviews of Continental mathematics in the *Monthly review*: in particular, I attribute to him the long piece on … Lagrange's *Functions*. (1990, 264)

Continuing the report of his travels with Wood and Woodhouse, Dupin was able to visit Cambridge's scientific facilities. From there, he went to Hull, where the engineers 'allowed him to copy the design of several fine cranes of wrought iron and to make detailed notes about the docks, which make this port one of the most interesting after London and Liverpool' (ASHM CC7 779). This does not mean, of course, that Dupin was permitted to see works that were considered to be of greater importance.

He went on to Newcastle, where he had seen 'a tower some 40 metres high for the production of lead balls' and 'the fine silver-gilt service, intended for the mine-owners to demonstrate their gratitude to Sir Humphrey Davy for his safety-lamp'. In light of such gestures of recognition, he added, 'a people deserves new discoveries ever to increase its industry, wealth, power and glory'.

Edinburgh and Leith, he noted, offered

> a mass of objects important to the sciences, literature and the arts; Edinburgh is the Athens of the North and the Scottish people combine the urbanity of the Greeks with the hospitality of the Arabs. The engineers acquired for me the plans of the works for the port of Leith where they are building docks, the form of construction of the warehouses, etc.

Dupin then crossed the 'bay of Firth', as he called the Firth of Forth, then that of the Tay, finally arriving in Dundee,

> where they are digging a new port, partly excavated in the rock. The engineer of this port, took me, not only to the linen spinning-mills, but after my departure, wrote to me to offer to draw me the design of the mechanism of one of his spinning machines, which had particularly attracted my attention.

Still according to his report (ASHM, CC7 779) from Dundee, Dupin went to Arbroath to visit the famous Bell-Rock lighthouse, twelve miles out to sea. From the coast, he was able to observe the red and white lights. It has been seen, in chapter 9.2, that he bought some red lenses before his departure from England; his friend Sganzin had asked him particularly to study the lighthouse because the red light lasted much longer than the white one.

Proceeding to Aberdeen, he met one 'Dr Gregory'—probably a descendant of the astronomer and mathematician James Gregory (1638–1675), who was on his way back from the Shetlands. Gregory took Dupin to an observatory 'as rich in instruments as it was poor in observations', and then to a leather and clog vendor who made barometers, thermometers, and large telescopes in his leisure time. To Dupin, the scientific instruments were much more interesting than the clogs, because by raising the telescopes up to the skylight, 'the sun and moon could be seen as well as in the most advanced observatory'.

From Aberdeen, Dupin went to Inverness, then Fort Augustus, and finally along the Caledonian Canal to Fort William. Telford's canal he described as 'an immense work'. It links the east and west coasts of Scotland, connecting Loch Lochy, Loch Oik, Loch Ness, and Loch Dochfour, and it enabled large ships to cross Scotland without having to negotiate the dangers of going around the coast via the north of Scotland and the Orkneys. Sadly, it never really paid for itself; steam power replaced sails, and the steamships were much better able to cope with rough seas. The locks were of wrought iron, and Dupin described the gates as weighing '25 to 26 tons', although he noted that they were 'opened and closed with the greatest of ease by just two men'. He added, 'I have taken a host of details of the structure of these gates'.

Next on his itinerary came the Glasgow and the Carron foundries, of which more details will be given in chapter 10. By the time he described the foundries, Dupin was nearing the end of his preliminary report, and he declared with enthusiasm,

> If I could, in such a brief exposé, give just a slight idea of all the institutions, the work in progress since a few years ago for the prosperity of Scotland, I would have presented one of the most perfect pictures to stir the admiration of all men and to occupy the minds of sages. It is beautiful to see a people using its labour, its dedication and its genius to master an unkind mother nature, to overcome the climate, to bring even sterility to fertility and, mindful of the riches of the soul as well as those of the senses, to cause to flourish together agriculture, trade and industry, ethics, religion and freedom.

If one overlooks the exaggerated writing style that was typical of the period, it is apparent that Dupin was highly delighted with his journey, which was a revelation to him, and with the welcome that had been accorded by the Scots to a Frenchman.

He then described his return to England via the west coast, revisiting Liverpool, which he had seen on his first trip. He came to it with renewed interest because, after a year's absence, he found 'a thousand new things'. Huge docks, which had been half-built the previous year, were now already complete, 'full of water and covered with ships'. He again visited the rope manufacturers, with their patented systems, and a manufacturer of steam pumps and high compression pumps. In Chester, he commented on the court of justice and its combined prison as 'the most beautiful of its kind to be seen in the whole of Europe'.

About his visit to Wales, Dupin waxed lyrical. He described the aqueducts of the Ellesmere Canal and the traffic on it, and then his journey to Llangollen:

> After a long and difficult walk, I entered the valley on a beautiful autumn evening, just before sunset. Never did I see a more imposing view. Amidst the lush vegetation still in all its freshness, the fires of the forges and the lime kilns belching their whirlwinds of flame and smoke, villages and factories … I remained as if in ecstasy.

Reading this, it is difficult to understand why Dupin's father could credit him with no literary talent.

Dupin then spoke of Thomas Telford, the engineer who was responsible for what he called the 'Ponte Cysille' aqueduct. It is hardly surprising that Dupin had difficulty with this name—the Pontcysyllte aqueduct on the Llangollen Canal at Chirk. Telford apparently had given him the design for it and then recommended him to the foundry which had produced parts for the aqueduct as well as the iron parts for the great lock gates on the Caledonian Canal.

Dupin then went to Holyhead, where he admired 'a magnificent breakwater of very great length'. This enabled big ships to enter the port at high or low tide, whatever the direction of the wind. At Holyhead, he embarked for Ireland and gave a brief description of Dublin. He found that in 'its population, grandeur, the beauty of its buildings, [and] the regularity of the modern part', it was 'one of the most remarkable cities of the British Empire'. New docks were being built beside 'the magnificent customs buildings; they will be vast and regular: the first stone was laid in the summer of 1817'. He added regretfully that he would have loved to tour Ireland as he had toured Great Britain, but the fall he had suffered (see chapter 9.2) prevented this.

The report ended with thanks and mainly favourable remarks to Dupin's British hosts:

> Considering all the wrong that the British administration has done and is still doing to us, it will perhaps astonish that I have only praise for the people I have met and the things I have seen. Nowhere in Great Britain did I see traces of the too long state of hostility which was pushed to excess at the height of the last war. Everywhere, I found kindness, hospitality and almost always generosity of communication, particularly among the engineers … Except in London, and then amongst very few people.

Dupin did not miss the opportunity to snipe at the French administration over his lack of funds:

> If I had been less severely limited by my small amount of funds, I would certainly have obtained greater results. But, restricted as I was … I may venture to believe that I have done everything that a patient, active and determined man could do … I had to struggle with the obstacles of nature, making difficult climbs, of wild and mountainous lands. I suffered all the inclemencies of a cruel sea, a harsh sky, hunger and thirst …, and all with pleasure, in the hope that one day all these discomforts would be useful to my homeland.

Nearing the end of his report, he made particular mention of the Andersonian Institution: 'The Anderson Institution [*sic*] has produced amazing results. It is a wonderful thing to see today, in many Glasgow workshops, simple workmen possess and develop when necessary, the principles of their social harmony'.

Dupin completed his narrative with thanks to the French people who had helped him in England; Séguier, the consul general; the consuls in

Dublin, Liverpool, and Edinburgh; and the René Eustache d'Osmond, comte de Boitron, the French ambassador to Britain from 1816 to 1819, who had enabled him to go to the military arsenals of the British navy on his first visit and had requested the prolongation of his leave on his second visit.

This was Dupin's draft report of his second visit. Much more detail was to be provided in the published versions (1819a and 1819b) particularly about his meeting with Andrew Ure—an encounter which was to have a profound effect on his future career.

However, overwhelmed by the welcome he had received and the success of his visit, he failed to see the negative aspects of British industrial progress and to put the new features such as steam engines, hydraulic presses, and iron roofs in context. Many innovations had disastrous consequences and were accompanied by serious accidents and deaths. At the manufactory of Henry Maudslay,[6] a man who was much admired by Dupin, the iron roof of the factory collapsed. However, Dupin saw only the successes of Maudslay's innovations (see Werner 2010, 201).

Alexander Galloway, who had a large manufactory in London employing eighty men, visited France in 1818 and again in 1823 to view the public exhibition in the Louvre. He was, according to Werner,

> astonished at the progress of French manufactures, particularly in 'bar and sheet iron, sheet steel, copper of every class, together with brass' ... British engineers, such as Donkin,[7] involved with French inventions were not investigated by Dupin. All he did was to lament how 'the most ingenious mechanics thus carry to a foreign country the treasure of their industry'. (Werner 2010, 206; his quotation is from the *First Report of the Select Committee on artizans and machinery, British Parliamentary Papers*, 1824)

In the next chapters, I shall study the official accounts of all of Dupin's visits. They would anger the French authorities because he said very little that was negative about British progress—probably because he did not see it. His somewhat naïf enthusiasm led him to believe that all doors had been opened to him.

ENDNOTES

1. Armand-Louis-Maurice Séguier (1770–1831), a diplomat and descendant of the famous chancellor Pierre Séguier (1588–1672).
2. Maria Agnesi (1718–1799), an Italian, published on differential and integral calculus in her 1848 *Analytical Institutions*, one of the first and most complete works on finite and infinitesimal analysis. The book became a model of clarity and was widely translated and used as a textbook.
3. Marie-Sophie Germain (1776–1831), one of the first self-taught French women mathematicians. Known for the arithmetical theorem which bears her name, for her exchanges with Carl Friedrich Gauss, and for her work on flexibility.
4. Sophie Kovalevskaya (1850–1891), the first professional Russian woman mathematician. She was awarded the Bordin prize at the Paris Academy of Sciences in 1888.
5. Ada Byron Lovelace (1815–1852), considered to be the first programmer in history. Her work assumed importance with the advent of computer science.
6. Henry Maudslay (1771–1831) was responsible for innovations in machines and steam engines.
7. B. Donkin (1768–1855), a manufacturer of paper and printing machinery.

CHAPTER 10

DUPIN'S PUBLICATIONS ON GREAT BRITAIN UNTIL 1824, RESULTING FROM HIS TRAVELS

10.1. THE *TABLEAU DE L'ARCHITECTURE NAVALE AUX 18E ET 19E SIÈCLES* (1815C)

As shown in chapter 6.2, the first part of the *Tableau de l'architecture navale aux 18e et 19e siècles*, drafted in Toulon, completed around September 1814 and published in 1815 (1815c), was the incentive for Dupin's visits to Britain. His two first journeys (1816–1817 and 1817–1818) and his tours of various arsenals enabled him to complete the work in Dunkirk after his return from the second visit. He had been greatly assisted by influential contacts and warm recommendations. The first part of his *Tableau*, having received an enthusiastic report from the directors of the Ports et arsenaux, was then submitted to the mathematical sciences class of the Institut in 1816. The second part was addressed to the Academy of Sciences, which entrusted to Sané the task of reporting on 'a work of M. Dupin concerning … naval architecture in England' (ASHM CC7 779; letter of 9 November 1818 from Sané to the minister for the navy and the colonies).

Dupin described his aims in the work thus:

> Architecture is not just the art of creating, drawing and building the walls …; it is the art of shaping and coordinating all the objects and all the parts of a construction. I may be at variance in that with commonly accepted ideas, but I believe I return to first principles. Thus, I see as included in naval architecture, masts, sails, rigging, installation and even lading, these accessories whose aim is in the best placing of the objects which the vessel must contain.[1]
>
> The first part presents naval architecture limited to the description of the most complete ship … I chose as the basis of my descriptions the three-decked vessel of M. Sané.
>
> The second part, describes every type of vessel from a first class man of war to the simplest barge. (1818b, 322–324)

The reports of both parts were brought to the attention of the minister for the navy and contributed to Dupin's advancement in that service. Included with his documents for his election to the academy, which were submitted on 6 December 1818, the reports were much to Dupin's advantage; he was elected on 28 September 1818. In addition, in order to be elected, his residence in Paris was compulsory. Dupin could not have hoped for more.

10.2. *Mémoires sur la marine et les ponts-et-chaussées* [*sic*] *de France et d'Angleterre* (1818b): The First Visit

Mémoires sur la marine et les ponts-et-chaussées [*sic*] *de France et d'Angleterre*[2] was dedicated to Prony, the former professor at the Ecole polytechnique and the director of the Ecole des ponts et chaussées.

Dupin's florid writing style was typical of the period, but the preface contains an interesting point:

> If, less obsessed by the only glory of being a pure and deep theoretician, each one of our savants were prepared, as you did,[3] to come down from the heights of his meditations, and not to scorn applications, encourage and respect them, we should soon have the honour of being in the front rank of nations where science and industry are the unfailing source of wealth, strength and grandeur. (1818b)

Dupin was constantly concerned with the application of theory, and this would inspire his lectures at the Conservatoire des arts et métiers.

He described the first two memoirs as

> a very rapid survey of the main perfections of public works in England, Scotland and Ireland … To expand and detail the idea of the progress of the arts in England, I describe, in two special memoirs, the operations which seemed to me to be the most remarkable in the construction of the Plymouth breakwater and the Caledonian Canal.

In his *Mémoires sur la marine*, he repeated points from his 1818 manuscript report (ASHM CC7 779), which was really just a draft. Now he was producing a detailed report of the first visit, which occupied the first fifty pages of the volume.

He explained that he had gone to Britain to complete his *Tableau de l'architecture navale*, a work he had begun in 1813 and 'pursued through all the obstacles of a difficult life and a disastrous period'. He referred to the commission which had acclaimed his work and declared that Dupin should conduct a survey of the main ports of France and abroad. He then recalled the work on the strength of timber he had carried out in Corfu and continued in Toulon and Dunkirk, which had prompted similar experiments in Great Britain. With regard to the British experiments, he added,

> Here is what Professor Barlow, author of the experiments I describe, had to say:
>
> … And it was not till after these had been completed, that I was favoured with M. Dupin's Memoir by Dr. Gregory, who had

> then recently received it from Paris, and in the perusal of which I felt great satisfaction, by finding my results verified by those of the above gentleman. A more recent course of experiments conducted with great accuracy and ability, by M. le Chevalier Dupin, has been published in the tenth volume of the 'Journal de l'Ecole Polytechnique', relating principally to the deflection of timber under various circumstances, and to which several references will be found in the following pages. ... I have since had the honour of several personal interviews with M. Dupin, who was present while I performed some experiments, and who has himself continued to pursue the subject with great attention and ability ... A more minute account of M. Dupin's results will be found in part III Barlow's [Barlow 1817] on strength and stress of timber London 1817. (Dupin 1818b, xi–xii)

Another reference to Gregory (see chapter 9.4) may be noted here.

After the introduction, Dupin went on to 'a succinct account of the first journey in the ports of England presented to the Academy of Sciences of the *Institut de France* on Monday, 26 January 1818' (traditionally, the academy met on Mondays). He found London to be the greatest trading port in England, a thriving centre of industry for the navy, and the focus of operations for the Royal Navy. He explained the main differences between the English and French ports, particularly in terms of the structure and construction of the dry and wet docks. He was struck by the constant use of the steam engine for hydraulic work in London as well as the presage of railways:

> Digging, the transport of stone, sand, lime, is all carried out by small four-wheeled carts drawn by a single horse on iron roads. These roads are composed of elements which can be placed or moved with the greatest of ease. It is to them that England owes some of her wealth. Never, without them, coal, minerals, raw materials of all kinds could have been transported great distances and with almost nil expenditure ... Underwater excavations, when the bottom is muddy and sandy, are carried out by pumps established on boats and moved by steam power ... A dredging boat, with a steam engine to work the dredging scoops, is taken to every

> part of the dock where it is necessary to use it. The mud-barges, as they are loaded with mud by the pump, are placed under another pump, on the edge of the dock, and moved by a steam engine fixed on the bank. (Dupin, 1818b)

Dupin then proceeded to other questions of metalworking, rope and cables, and provisions for the teams:

> The British Government would see it as not only as barbaric, but as madness to economise at the expense of the health of the men who give their strength and their lives to her defence … When I report that the teams lunch on chocolate, this will, perhaps, provoke laughter in the shallow observer, but it will strike a chord among those men who, mindful of the lessons of Hannibal, are aware of the extent to which physical strength in addition to strength of mind can decide the outcome of battles.

He returned to the question of timber, a constant preoccupation of visitors to Great Britain (see Bradley 2005): 'Some two or three years ago, great experiments were conducted at Woolwich on the impregnation of wood with a mineral solution to preserve it from perishing. To this effect, they used a high-pressure steam apparatus'.

It is interesting that Dupin now seemed to express a different attitude towards Seppings from that previously expressed (see chapter 7.4):

> It was at Chatham that Mr Seppings implemented, for the first time, the improvements in ship construction for which he is responsible. To achieve this he had to overcome the ancient prejudices of our ancestors … I have tried to make known in France the genuine advantages of the Seppings system. I found more obstacles than he did and, less fortunate than he, I have not overcome them. Seppings triumphed over the opposition, and rendered services to the British navy better than it had ever received before. (Dupin 1818b, 32–33)

Dupin then described a visit to the celebrated ship *Bellerophon*, a ship that was perhaps notorious as far as the French were concerned for having been so instrumental in the downfall of Napoleon. She had fought at the battles of the Glorious First of June, the Nile, and Trafalgar and was one of the most famous ships of the Napoleonic Wars. To her crew, she was affectionately known as 'Billy Ruffian' (or 'Ruff'n'). On 15 July 1815, Napoleon surrendered to Captain Frederick Maitland of *Bellerophon* and remained aboard. On 4 August, the ship was ordered to sea again from Plymouth, to await the arrival of HMS *Northumberland*, which was designated to take Napoleon to exile on St. Helena (see Cordingly 2003).

At the time of Dupin's visit, *Bellerophon* had been converted into a prison ship, and he commented, 'In the equipment and furnishing of this prison ship, everything humanly possible has been done to make a floating prison bearable and comfortable'. In a footnote, he added a linguistic comment on the word 'comfort', spelled with an 'm' rather than the French 'n', explaining what was meant by an English word that was unknown in France:

> The words comfortable and comforts are amongst those expressions missing from the French language, although we have the derived words conforter and réconforter (with an 'n').
>
> Comfortable implies what is pleasant and useful, all that removes every irksome or repellent sensation in the satisfaction of our needs. Comforts thus render our existence free of unpleasant needs and feelings. Thus the English, a highly sensual people, create the happiness of their lives in having 'their four pleasures' in everything, and at all times use the words comfort and comfortable to describe all the things they possess or desire. (1818b)

Dupin was much impressed by the Portsmouth naval hospital, where patients could take hot water and steam baths:

> The water of these baths is steam-heated. The laundry and the driers are also worthy of note. When Howard,[4] the great humanitarian, visited for the first time, the prisons and hospitals of England, he painted an appalling picture of the sad state of the majority of these establishments; he had scarcely anything but praise for the Portsmouth [naval] hospital and, above all for that in Plymouth.

> So the navy was far ahead of the rest of society, in the help and care invented by the most beneficent of the arts, that of healing. (1818b, 41–42)

He added a footnote:

> English sailors love life's comforts and, in many respects, they enjoy them as much as the most wealthy citizens. The welfare they experience consoles them for the austerity of discipline, and they remain happily aboard for two or three years in a row, on the most dangerous and difficult voyages.

Life in the navy over the centuries has been described in a comprehensive history (Rodger 2004). In the early nineteenth century, a liberal strain of opinion 'urged that seamen be treated as intelligent adults and at least potential citizens'. Another feeling was that 'the education of the lower classes was of the utmost consequence to the permanent security of the empire, even for the sake of our navy'. The captain's object should be to command 'a cheerful and contented crew ... Seamen knew and valued the necessity of good discipline afloat, for the comfort and safety of all' (491–492).

There were inevitably dangers to health and morale attached to long sea journeys, but the men were given 'every opportunity to buy food and comforts for their messes. In foreign ports, men invariably bought fruit, vegetable and livestock' (Rodger 2004, 502). The harsh treatment that was often meted out on ships is well known, but one of many such anecdotes shows another side of naval life. According to one story, the captain of a ship

> was amused upon going round the ship last Sunday, to see a whole sheep roasting in the galley, stuffed with potatoes and onions. It seemed that the mess to which this belonged had bought it ... for a Christmas dinner: but ... they determined 'to have a good *blow-out* while they were all stout and hearty'. (Rodger 2004, 502)

The needs of British crews were catered for as far as was possible, and when the sailors could not go ashore, the jollyboats provided extra solace. As regards the Portsmouth hospital, this had already impressed the young French engineer Joseph Cachin in 1785. In the report of his visit to Britain, he referred to the 'fine hospital for sick and wounded sailors' (ENPC, ms 87).

10.3. The Second Visit

Dupin then turned to his second visit, where he had particular praise for the warmth of the Scots and their liberal approach to visitors. One meeting which was to have very important consequences for him was that with Andrew Ure at the Andersonian Institution. In Corfu, Dupin had already stressed the urgency for education in the sciences and mathematics as the key to prosperity and the theoretical means to attain the most perfect possible practical results.

> The main professor at the Anderson [*sic*] Institution, Doctor Ure, known for his kindness to foreigners and particularly to the French, himself took me to all the important manufactures, of which a large number are today managed by the students … To give … an example of the education of the simple workers of Glasgow, I shall quote the case of two bakers, brothers, who in the interval between one baking and the next, occupy themselves with making machines and instruments of physics … They have built a small apparatus for lighting their shops and apartments … The young men are well acquainted with the principles of physics and mathematics of the instruments and machines they create. One day they will leave their profession to pursue, and I venture to predict it, with success, the natural sciences. (1818b)

Dupin's implication was that France could learn a great deal from this example if it were to emulate British industrial prowess.

10.4. Andrew Ure (1778–1857)

Who was Andrew Ure? He had held the chair of natural philosophy at the Andersonian Institution for more than ten years when Dupin met him, and he was a most interesting character. He had studied medicine and served briefly as an army surgeon. In 1804, he replaced George Birkbeck[5] as the professor of natural philosophy at the Andersonian Institution. In 1808, he founded the Garnet Hill observatory and in 1811 was elected a fellow of the Royal Astronomical Society. He published widely, with a particular interest in the welfare of the working class, and in 1839, he produced his *Dictionary of Arts, Manufactures and Mines*. In the 1853 fourth edition of this work, it is clear that he envisaged a utopia where the work of the artisan would be replaced by that of the automated factory (see Mertens 2006).

In 1818, Ure revealed that he had been conducting experiments on a man who had been executed by hanging. He claimed that by stimulating the phrenic nerve, life could be restored in cases of suffocation, drowning, or hanging. It is said that this influenced Mary Shelley when she wrote *Frankenstein* and indeed, that Ure was the model for the main character. The Shelleys took a keen interest in the scientific developments of their day, and Percy Bysshe Shelley was particularly enthusiastic about experiments such as Ure's. While he was trying out electrotherapy, he managed to electrocute the family cat (see Bakewell 2000).

According to the obituary in the *Gentleman's Magazine* (February 1857), 'Dr. Andrew Ure, FRS, FRE [Fellow of the Royal Society of London, Fellow of the Royal Society of Edinburgh], a chemist and a contemporary of [Humphry] Davy, [William Hyde] Wollaston, and [Louis Joseph] Gay Lussac, was distinguished by his success in the application of chemistry to the arts and manufactures and unrivalled both as a teacher and a writer'. The obituary continued,

> It was in consequence of an official report by M. (afterwards Baron) Charles Dupin to the French Government, on the influence of Dr Ure's tuition on the manufactures of Glasgow, that new courses of lectures on the same plan were instituted at the *Conservatoire des arts et métiers* in Paris, one course, '*Sur la Mechanique Appliquée*' under M. Charles Dupin, and another '*Sur la Chimie Appliquée*', under M. Clément Désormes.[6] Dr Ure had then for many years superintended the practical school of

> chemistry at the Andersonian University [*sic*] of Glasgow. The fruits of this tuition will be best appreciated from the words of Baron Dupin's report: 'L'Institution Andersonienne a produit des résultats étonnants' [The Andersonian Institution has produced astonishing results] … This was in 1817. The teaching of the application of chemistry to the arts has since been introduced into almost every university, and into many public institutions.

It would seem that Dupin's interest in Ure was very much in the sphere of workers' education and Ure's achievements in that direction, probably much to the relief of the Dupin family cat.

10.5. Further Visits

After this momentous visit to Ure, Dupin visited the Carron Foundry and related an interesting story. The foundry's doors had apparently remained closed to Prince Nicholas of Russia, but Dupin was

> recommended as an amateur … Are you a salesman?—No! and with no interest in any business at all! Why are you travelling?—As a friend of the sciences and the arts, for my education. Let this gentleman come in (Dupin 1818b)

This contrasts sharply with the reception that was accorded to Faujas de Saint-Fond (1770–1842), the French geologist and vulcanologist, at the end of the eighteenth century; he was unable to gain admission to a single cotton mill and at Carron was allowed to see only the ironworks and not the cannon factory. He was 'not at liberty to take notes in writing and had to record his experiences until late into the night' (Geikie 1907, 1:xvi).

Dupin was well received by such great men as George Rennie and Thomas Telford, and he was particularly impressed by his meeting with James Watt in Glasgow:

> It is with respect mingled with admiration that I saw this handsome old gentleman of 83 years, retaining the vigour of his mind

> as well as his physical strength, and apprising me of a mass of details regarding the progress of British industry, of which, more than any other inventor, he was responsible for the acceleration. (Dupin 1818b, 71–72)

After a pleasant and fruitful visit to Scotland, Dupin proceeded to Dublin, about which he was far less enthusiastic. He claimed that foreigners were not admitted to places there as they were in Scotland:

> All the literary establishments are stamped with the seal of meanness ... No foreigner is admitted to enjoy the scientific collections, nor to read the scientific periodicals in the great reading rooms, nor to consult the books of the libraries.

However, his memory may have been somewhat soured by the injuries he suffered during his visit. Two letters relating to the *Mémoires* have been found, the first to the *ponts et chaussées* engineer C. L. M. H. Navier,[7] written from Dunkirk on 11 June 1818:

> I have just received a copy of the 'Mémoires sur la Marine et les Ponts-et-Chaussées' which I have just published. I would wish that this work might meet with your approval. You could do me a great favour which I venture to ask of you as a former schoolmate and, on a similar occasion, I would do the same for you with all my heart if I were able to. It would be to write, on the part 'Travaux maritimes et ponts' of this work, an analysis ... We would publish it in the Moniteur and the engineers of your Corps, seeing your approbation which would give them confidence would lose the prejudices which they would naturally experience towards a constructor who talks docks, bridges and canals ... I have dedicated my work to M. de Prony, our mutual friend; that should be, for you, yet another motive for taking an interest in this work. (ADN; PJ 73)

Dupin had returned to Dunkirk on 11 February 1818, and his *Mémoires sur la Marine et les ponts-et-chaussées* appeared in June, within the four-month period he had promised the minister for the navy.

The second letter is addressed to the British Museum librarian. It was written from Dunkirk and headed L'ingénieur de la Marine en chef (engineer in chief of the navy), so it was very much an official communication:

> I am honoured to send you three copies of my *Mémoires sur la Marine et les Ponts-et-Chaussées*; the first is for you, the second for Dr Brewster[8] in Edinburgh and the third for Mr Jeffries,[9] Edinburgh lawyer and editor of the Edinburgh Review.
>
> I hope you will be so good as to forward these copies to their addresses. Moreover, this will give you the opportunity to become acquainted with men of rare worth and high reputation. (ADN; PJ 74)

10.6. Other Publications of This Period

Dupin published several other works during the years 1820 to 1824, including *La force militaire de la Grande-Bretagne* (The military strength of Great Britain; 1820) and the *Force navale de la Grande Bretagne* (The naval strength of Great Britain). He was becoming increasingly interested in economics and administration, as is apparent from his *Système de l'administration britannique en 1822* (The British system of administration), published in 1823. His interest is even more in evidence in two other works, *Du commerce et de ses travaux publics en Angleterre et en France* (Trade and public works in England and France) and the *Voyages dans la Grande Bretagne* (Travels in Great Britain; 1825g).

From 1816 onwards, Dupin was bringing an end to his scientific activity and his work as an engineer. Gradually, he was becoming an economist and an educator. He had been an educator at the Ionian Academy, and in 1819, was appointed to a chair at the Conservatoire des arts et métiers (more will be said of this in chapter 13). He was increasingly an economist after his study visits to Britain. The *Mémoire sur la structure des vaisseaux anglais* (1816–1817) was the work of an engineer, but his interests were becoming less and less technical.

Although I shall now look at these works in connection with Dupin's study visits to Britain, I shall return to them when I consider Dupin as an economist. The *Système de l'administration britannique en 1822* was a study of finance, industry, commerce, and navigation. Among Dupin's titles as an author, he was now able to add member of the *comité consultatif des arts et manufactures* (the consultative committee of arts and manufacturing). In the introduction to the *Système*, he announced his aim to 'present a summary of the methods employed by the British administration in the short space of one year, to improve the fortune of the State and the welfare of its citizens'. By this time, Dupin had visited England five times. He wrote,

> In England, the stability of the Ministry is guaranteed by the general opinion of enlightened citizens. The safest means the Administration has to secure the support for its principles as well as its actions, of both citizen and legislator, is to explain frankly and clearly the path it has pursued and the fruits it has harvested … France, amongst all continental powers must be the first to see its Administration, not only equal, but surpass in moral courage the Administration of England and say to all her citizens: That is the spirit in which we have had to govern you during the year which has just passed. Such were our resources and our difficulties. Such are our results and our triumphs … We have omitted discussion of transitory events, political interests or party hostilities like mundane friendships which, according to the eloquence of Bossuet, change and vanish with the interests and the years. (1823b)

The first chapter is on finance, and Dupin claimed that the British administration said that the nation owed gratitude to its ministers in proportion to the services they had rendered to their country. They would obtain their gratitude by demonstrating that they had administered the national finances with just consideration of the people's welfare and the security of the state. In the chapter on navigation, Dupin stressed the importance of shipping to industry and prosperity. The shipping industry was the first to suffer from the abrupt passage from war to peacetime. Ships that had been built for battle had to be converted for trade purposes, but they then fulfilled a new purpose and needed regular maintenance. New ships were also increasingly necessary, so the industry was thriving. Dupin added statistics about the merchant ships that were built in 1820, 1821, and 1822 and their tonnage—perhaps a far cry from his work on stability and wood. As for internal trade, Dupin found this to be even more satisfactory than shipping and overseas trade. The most thriving manufactures he had seen were wool, cotton, linen, silk, hardware, and cutlery, he observed, but there was extraordinary activity and an aim for perfection in every type of British industry. Referring to silk production, he explained that he believed 'the English are threatening with imminent ruin one of the most brilliant, lucrative and important of our national industries'.

The main points arising from these chapters are Dupin's admiration for the openness of the British administration regarding results, good or bad, and for the management which he believed the French should emulate. However, this attitude was hardly likely to endear him to the French authorities. As it was, in 1823, the Villèle[10] government was stringent, and liberal laws, for instance relating to the press, had been repealed in 1822. As will be seen, Dupin was to denounce Villèle's reactionary policies in 1827.

The discourse *Du commerce et de ses travaux publics en Angleterre et en France* (Trade and public works in England and France; 1823a) was given to a session of the Academy of Sciences, of which Dupin had now been a member for five years. This speech was an abridged version of what was to be the third part of his *Voyages dans la Grande-Bretagne* and was entitled 'Force commerciale' (Commercial power). In it, Dupin revealed the object of the future third part of his *Voyages en Grande-Bretagne*: the reasons for Britain's supremacy in trade, so that France and other countries could benefit from the lessons that were involved. He claimed that it was more useful to know how other nations had overtaken France rather than being satisfied with the country's own achievements.

He felt obliged to defend himself against the accusation of Anglomania of which he was sometimes the victim. He recalled that in 1817 at a meeting of the Royal Society, he had claimed for 'our engineers and our savants, the prime honour of the most beautiful achievements in an art on which the British Empire depends in part for the security and power, naval architecture' (1823a). He also reminded his readers that when 'in the House of Lords of the British Parliament, Lord Stanhope proposed to prolong the scourge of the occupation of our land', he had championed the cause of France (see chapter 9.3). In spite of such protestations, Dupin was flatly accused at this time of being too liberal in his attitude towards England. In particular, as will be seen in the following sections, the publication of his *Force militaire de la Grande-Bretagne* in 1820 greatly angered the French government, which saw it as being far too favourable to Britain. The work was referred by the ministry for the navy to the Council of Ministers and was censored; Dupin's career suffered as a result and his works continued to be censored for at least two years.

He continued in his discourse to speak of Britain's great empire and powerful navy, which even in peacetime could set sail at the first sign of any threat to British territory. He compared Britain to ancient Rome: 'The British Empire has then, of itself, a principle of resistance which was lacking in the Roman Empire: this principle is that of commercial power'. Trying to find the reasons for this power, he added,

> It is not only courage, intelligence and activity, it is wisdom and economy, and above all the integrity of the industrious man which maintain the superiority of the production and trade of his country. If ever, in the British Isles, the useful citizen lost these virtues, you may be certain that for England, as for any other country, in spite of the protection of the most formidable naval fleets, in spite of the foresight and the aid of the most widespread diplomacy and the most profound policy, soon the ships of a declining trade, repulsed from all shores, would disappear from the seas which they cover today with the treasures of the universe. (1823a)

He then commented on the British character:

> Observe this character, inspiring impetus to the thought as well as action of the individual, arousing an irresistible fervour, insatiable,

> to overtake all rivals, and above all to crush the foreigner by a competitive spirit at once personal and national! And what means to attain this end! Cool, continuous and methodical activity.

On the material causes for Britain's prosperity, he added,

> As for the material causes, we would place first the public road network and the establishments which facilitate the transport and delivery, in the interior and in the coastal regions, the very art of transport and exchange; and the creation of industrial products, which are the material of these exchanges.

He stated that it was this type of order that should serve as an example for France—that the interior above all should be stimulated and rallied with an energy that would bring the country to her rightful place.

10.7. *Voyages dans la Grande-Bretagne* (1825)

As already indicated, *Voyages dans la Grande-Bretagne* (1825g) was a compilation of the material of the *Force militaire* (the first edition of which was published in 1820 two volumes, with maps) and the *Force navale* (the first edition of which was published in 1821 in two volumes. It also includes the *Force commerciale*, of which Dupin had given a foretaste in his speech 'Du commerce et de ses travaux publics en Angleterre et en France'. In that speech, he announced that the *Force commerciale* would comprise a first section on public works, bridges and highways, trading ports, and so on, and a second section on transport on highways, canals, rivers, and the seas as well as maritime trade.

The work was published in 1825 with its full title, *Voyages dans la Grande-Bretagne, entrepris relativement aux services publics de la Guerre, de la Marine et des Ponts-et-Chaussées, au commerce et à l'industrie depuis 1816* (Travels in Great Britain, undertaken since 1816 for the War Office, the Navy and the Bridges and highways, and trade and industry). To his name, Dupin appended a complete list of all the honours he had received. He dedicated the work to 'His Royal Highness, Monseigneur le Dauphin, admiral of France', this being the Duc d'Angoulême, the son of Charles X. Interestingly, when the Ecole polytechnique was reorganized by the restored monarchy in 1816, the king declared, 'The Royal Ecole polytechnique will henceforth be in the protection of our beloved nephew, the Duc d'Angoulême' (Fourcy 1828, 337–338).

Dupin's pompous and obsequious style is typical of the period: 'I propose to bring together in one work the collection of fact studied amongst a people famous for its prosperity, so as to call our homeland to similar and even greater wealth'.

However, he trod on a few toes:

> As soon as an improvement is proposed, moderate, tardy, even minor, some interests are upset, some self-esteem believes itself offended and consequently is so; and innumerable obstacles are put in the way of future advantage, even to the benefit where fertile seeds seem to promise the finest fruits. (Dupin, 1825g)

There is no doubt that Dupin was taking a risk in writing this book; in each of its three parts, he sought to inform the French of Britain's superiority in many areas, but at the same time, he wanted to warn them of the danger to France of stagnation or even progress that was too slow. He was inevitably labelled an Anglophile, even though he defended himself many times and with good arguments against this accusation. He clearly stated his admiration for great French achievements.

At the beginning of the nineteenth century, Dupin observed, England was developing rapidly, with canals for transport from the mines, for navigation, and for the distribution of water to the people. Gas was lighting the streets, ports and docks were being built, and newly established breakwaters, jetties, and lighthouses were increasing security. However, he added, in defence of France,

> It is not sufficiently well known amongst us, that the magnificent roads, opened from the capital to the main points of our frontiers,

> were created for the most part under the ministry of the Duc de Choiseul,[11] one of the first statesmen of the 18th century. (1825g)

At the end of his introduction, Dupin thanked all those people who had assisted him in his work, including famous English savants and 'English officers of all grades and all services'. He then added great French officers, a marshal and generals, and 'above all, the officers of the artillery and navy'. Finally, he named three of his former classmates: Parizot (whose first name is not to be found; he was to be referred to again in Dupin's *Rapport du Jury* of 1836), Henri Joseph Paixhans,[12] and his old friend Augoyat (see chapter 3.1). For them, he said, 'I found amongst these officers, friends of my youth; they helped me, illuminated me, corrected with fraternal zeal and affection, one of the fine outcomes of our *polytechnique* education'.

He referred to the arms that were 'produced in workshops directed by Colonel Parizot with a rare talent'. Once again, he thanked Séguier (see chapter 9.1) for his 'observations and manuscripts'. As has been seen, this work, a collection of his publications of three and four years before, clearly illustrates Dupin's progression and his change of direction from engineering to economics.

10.8. Espionage?

As shown in chapter 7, espionage was rife in this period, but was Dupin also involved in it? The word itself is fashionable among historians, and there is a temptation to attribute nefarious intentions to innocent activities. It was perfectly natural for visitors to travel to Britain given that after 1815, new processes and inventions were being used there and also that travel to the islands had for so long been impossible. There were, of course, many examples of visitors with clandestine intentions, such as Dufaud or de Gallois (see chapter 7.1), who came on behalf of companies to discover the secrets of rival enterprises (see Bradley 1995). On his return home, Louis de Gallois presented a paper to the Académie royale des sciences entitled 'Mines, usines et chemins de fer'

(Mines, factories and railways), in which it was clear that he had been intent on gleaning all of the information that he could during his visit. Dupin's travels, though they were equally motivated by a desire for knowledge, had a clear intention.

Dupin referred to de Gallois in chapter 6 of the third part of the first volume of his *Voyages dans la Grande-Bretagne*, a chapter entitled 'Routes-ornières en bois et en fer des rail-roads d'Angleterre' (Grooved wooden and iron rails of English railways). He mentioned what he described as 'a very interesting paper by de Gallois on railways'. Dupin called the iron rails that were used in Britain to transport goods 'grooved roads', but he sometimes used the term 'railways'. He suggested that

> on certain very steep roads in our big towns, for example in several descents from the Montagne Saint-Geneviève in Paris, would it not be possible to place beside the road similar flat grooves? They could be used by carriages going up, whilst those going down would be driven on the ordinary surface.

He foresaw the possibility of passenger transport.

Charles Dupin was a state official, a naval engineer. He claimed that the main reason for his visits to Britain was the completion of his *Tableau de l'architecture navale*. It may also be noted that he was to visit French ports as well as foreign ones. He gave the impression that he was above suspicion and completely honest in his approaches. However, a doubt may linger. The fact remains that the British were very sure of their supremacy, and visitors meant potential orders from their manufacturers. British manufacturers were free to register import patents, but they rarely did so, John Walters being one of the exceptions (see chapter 7.4 and PJ 48). It is probable that although many industrialists appeared to be completely open, they were keeping important secrets to themselves and therefore did not feel the need for import patents. The export of British machinery and tools was restricted, but specifications, descriptions, and drawings were freely available. Alexander Galloway commented that this gave foreigners

> all the advantage of our knowledge, and … the means of fabricating all we know, with as much readiness as any native of this country can possess; and in many instances, patent machines are known sooner in France than in this country. (*Six Reports from the Select Committee on Artizans* [*sic*] *and Machinery*, 23 February–21 May 1824, 19, British Parliamentary papers)

According to Werner, Galloway knew Dupin and respected him as being 'most able' with 'a great mechanical knowledge', but he revealed that Dupin had approached him to supply him with drawings (2006, 206). Werner commented,

> What would Dupin have done with them? No doubt he would have circulated them to French engineers and manufacturers and deposited them in relevant archives where they could be easily consulted. Thus, the French would have been kept up-to-date with the latest British technological inventions and improvements. (2006, 206)

Whatever his motives, Dupin considered himself to be much indebted to his British hosts, and in his *Mémoires sur la marine* (1818b), he declared,

> The Royal Artillery and Military Engineering School, which used to be part of the artillery arsenal, is now separate from it. If I owed to Sir Joseph Banks, president of the Royal Society of London, only the acquaintance and friendship of Colonel Mudge, governor of this school, it would be enough to warrant my eternal gratitude to this patron of the sciences of Great Britain.

As for Scotland, he stated,

> It is easier to visit the establishments and workshops of Glasgow than those of any other city in the British Empire. The liberal spirit of the people extends as far as possible in a manufacturing society which must naturally fear, and try to prevent, not only the loss of its supremacy, but all foreign competition.

Dupin was honoured with the confidence of William Huskisson,[13] the British president of the Board of Trade and treasurer of the navy, who provided him with a great deal of information. In a letter sent from London in 1824 to Christophe de Chabrol, the secretary of state in the Navy Department, Prince de Polignac[14] wrote of Dupin,

> What I may assure you is that the immense work to which he devoted himself is so greatly appreciated in England that the members of the English ministry have several times told me that they there found information which they could not find elsewhere. (ASHM CC7 779; PJ 75)

In conclusion, after the fall of Napoleon I, a generally good collaboration in the industrial field was established between France and England—at least, according to Dupin in volume 1 of his *Voyages dans la Grande Bretagne*, (1825g) the *Constitution de l'armée*:

> We hardly benefited from the short peace following the Treaty of Amiens, to become acquainted with the new progress of the British in various branches of industry. The desperate war prevented for twelve years, new inventions and improvements from crossing the Channel to be propagated on our soil. But, since the end of that war, in different circumstances, a happy exchange of research and discoveries has been established between nations previously at war. French engineers have visited British establishments; British engineers have visited them in France. The narrow jealousy of governments has gradually yielded to fairer notions on contacts which hold, in themselves, the prosperity of nations.

ENDNOTES

1. Compare this with the work of Burgues de Missiessy (1798) and what he called 'la charge'—the entire sum of all the objects aboard a ship.
2. Names appear with hyphens and without them. During this time, there were inconsistencies.
3. This is possibly a reference to the Prony brake, or dynamometer (see Bradley 1998, 29); Prony always attached great importance to the application of theory.
4. John Howard, an English prison reformer. His 1777 *The State of Prisons* advocated improvements to enhance the physical and mental health of prisoners and the security and order of the prison.
5. George Birkbeck (1776–1841), a professor of natural philosophy at the Andersonian Institution and the first president of the London Mechanics' Institute, which became Birkbeck College.
6. N. Clément-Desormes (1770–1842), a physicist and successful industrialist. More will be said of Desormes's connection with the Conservatoire des arts et métiers.
7. C. L. M. H. Navier (1785–1835), an engineer in the Corps des Ponts et chaussées. In 1824, he was elected to the Academy of Sciences. He is renowned for his work in fluid mechanics and the Navier-Stokes equations.
8. Probably Sir David Brewster (1781–1868), a Scottish physicist who studied the polarization of light.
9. Francis Jeffrey (1773–1850). The *Edinburgh Review* was Whiggish in politics, in opposition to the London-based and Tory *Quarterly Review*.
10. Comte de J.-B. Villèle (1773–1854) headed the French government from 1821, representing the Ultraroyalists. He took the reins of public education, supported the church, and opposed the freedom of the press.
11. E. F. de Choiseul (1719–1785), the secretary of state for foreign affairs, war, and the navy from 1758 to 1777.
12. H. J. Paixhans (1783–1854) of the Corps royal de l'artillerie was an armaments and explosives expert.
13. William Huskisson (1770–1830). He was the world's first widely reported railway casualty, being run over and killed by George Stephenson's Rocket.
14. Prince J. de Polignac (1780–1847), the French ultraroyalist politician and French ambassador in London from 1823 to 1829.

Chapter 11

A Summary of the Rise of Charles Dupin to High Honours and the Start of a New Career

11.1. Dupin's Period of Change

In the previous chapters, I have studied the works of Dupin up to the year 1824. I shall now return to the years 1816 through 1819, because it was in this period that he began to change directions in his career. This transformation resulted from his travels in England, and his writings illustrate a progression of his interests from engineering to economics. Moreover, during these years, he won honours to which he had long aspired and which would considerably raise his social position.

On 1 September 1818, the minister for the navy and the colonies decided that Dupin would have to reside in Paris, fulfilling a desire that Dupin had nurtured since his return from Corfu (see PJ 27).

On 28 September 1818, he was elected to the Academy of Sciences. On 17 March 1819, he was promoted to the post of ingénieur de 3ème classe in the Génie maritime; previously he had been a sous-ingénieur (the grade of ingénieur de 3ème classe corresponded approximately to that of a frigate captain—a superior officer). On 25 November 1819, he was appointed a professor at the Conservatoire des arts et métiers, of which much more will be said.

Since the beginning of 1816—when Dupin had asked his friend Mathieu if one had to be 'a viscount, marquess or astronomer' to enter the Academy of Sciences (see chapter 6.7 and PJ 36)—and his unsuccessful campaign for the vacant seat of the deceased physicist de Rochon in April 1817, circumstances had begun to change for him. He strengthened his reputation through his two visits to Britain, his *Tableau de l'architecture navale*, and the *Mémoires sur la marine et les ponts-et-Chaussées*, enhancing what he had already acquired as early as 1813 with his *Développements de géométrie* and then with his *Structure des vaisseaux anglais* in 1816. It has been seen that in his impatience to be elected to the Académie des sciences, Dupin in 1817 had written to his elder brother that his future depended on it (see chapter 8.2 and PJ 58). In 1818, he took the seat that was vacated by Périer, the industrialist and inventor (see chapter 1.1). Dupin's life was indeed to be greatly affected by his election; he would rise rapidly to a higher position in the navy, his career as an educator and economist would soon take shape, and honours would fall almost automatically upon him.

Dupin's work as an educator and economist would finally lead him into a political career in 1827; his being a superior officer must certainly have hastened his progress. However, his father greeted his success with his usual sarcasm. In a letter of 26 March 1819 addressed to 'M. Dupin, ingénieur de la marine, rue de l'Enfer, no. 5', Dupin père congratulated his son in ambiguous terms that essentially nullified his compliments:

> I am rushing to congratulate myself along with you on your new grade. It is, for me and my family, another reason for loyalty to the king … It is he who made you a member of the Legion of

> Honour, the *Institut*, and superior officer of the *Génie maritime*. I see that you have taken care to consult several men of letters to indicate to you the faults in this plan, thought, style, etc., … But it is not enough to ask for opinions, you must take heed of them; it is not right that an academician be at once both heretical and schismatic in irreverence to, for example, authority and the *Académie française* where language is concerned. (DRB; PJ 76)

There would be no expressions of pleasure or pride from his father.

At the start of 1819, Charles Dupin's social status had greatly improved, although his ideas were upsetting the government of Louis XVIII. Fortunately, the king was more liberal than the people in his administration, and he lifted the prohibition on Dupin's works, which had been imposed when he published his *Force militaire de la Grande Bretagne* (see chapter 10.6). Around this time, 1819, Dupin was considering the introduction of popular education courses, such as he had seen in Glasgow, and the minister of the interior, E. Decazes (1780–1860), on the advice of his counsellors and the chemist J. A. Chaptal (1756–1832), created at the Conservatoire royal des arts et métiers a high school for the application of science to trade and industry (Haute Ecole d'application de la science au commerce et à l'industrie).

11.2. Appointment to the Conservatoire des arts et métiers and Further Changes for Dupin

With the end of the period of the Napoleonic Empire, France was faced with a new political and economic situation. The country was occupied and having to pay considerable compensation to its neighbours; moreover, it had been cut off from Britain, the main source of industrial innovations. Dupin had been among those people who took advantage of the peace to cross the Channel; another was Jean-Baptiste Say. Both men were to return with reports of Britain's prowess, implying severely adverse comparisons with the French situation. Industry must be reorganized and given the means to develop: this was their message. The question of manpower was primordial and immediate. To work steam engines, to install and manage weaving machines, French industrialists were obliged to import professional and technical workers. A national solution was therefore essential (see Grelon 1996, 40).

The courses at the conservatoire were useful, but according to Comberousse, 'they are unable to provide industry with skilled engineers equipped with the components indispensable to the new conditions in society' (1879, 7). The *écoles d'arts et métiers* in Châlons and Angers and the Ecole de maîtres-mineurs in Saint-Etienne were training supervisors and overseers. The teaching was essentially practical and not sufficiently advanced for future engineer-directors to be found there. The university faculties concentrated on the pure sciences and were no solution to the immediate problem. It was the public lectures on geometry and mechanics that were created by the initiative of Dupin and, at Metz, by General J. V. Poncelet[1] which had a focus on educating the working class. Since the restoration of the monarchy, Dupin had been proposing the institution of an *école polytechnique de l'industrie civile* (a polytechnic school for civil industry), in spite of strong protests from A. Lavallée[2] (see Mouchelet 1913).

In 1970, the French Ministry of Education produced a work entitled *1820–1970, cent cinquante ans de haute enseignement technique au Conservatoire des arts et métiers* (One hundred and fifty years of higher technical education at the Conservatoire des arts et métiers). On the role of Charles Dupin in the conservatoire, it comments,

> It would seem that it is to the *polytechnicien* Charles Dupin that we must attribute the initiative for teaching to introduce producers to the laws of physics governing the substance of their work and also the economic laws which determine the work itself … An interesting study by André Liesse [a political economist and statistician], tells us how, in Glasgow, Dupin discovered a school of sciences applied to the arts. It so greatly attracted his attention that on his return to France, he devoted a complete review of it in his *Mémoire sur la Marine et les Ponts-et Chaussées en Angleterre et en France*. And he concluded on this point expressing the desire that a similar institution be created. At that

> time we were under the ministry of Decazes … Decazes allied himself easily with Dupin's views and appointed a three-man commission to study the creation of higher technical education. (CNAM 1970, 20–21)

Among these three men was François Arago (see chapter 3.3), thirty-three years old, the youngest of the trio, who served as secretary. The other two were chemists, Louis Jacques Thénard (1777–1857), aged forty-two, who was very influential in the sphere of education, and Nicolas Clément-Desormes, a future professor, then aged just forty. Decazes's decree of 1819 stipulated that 'at the *Conservatoire* … public and free teaching will be provided for the application of the sciences to the industrial arts … This teaching will comprise three courses, mechanics, chemistry applied to the arts and industrial economics' (CNAM 1970, 21).

The first three professors—Charles Dupin; Nicolas Clément-Desormes (1778–1841), a manufacturer and physicist who is known for his work on lighting; and Jean-Baptiste Say (1767–1832), who is considered to be the first French classical economist—were also appointed by a decree dated 25 November 1819. That decree marks, in the history of the CNAM (the Conservatoire national des arts et métiers), the start of a new era to which Dupin would be inextricably linked.

At Dupin's funeral in 1873, General Morin commented,

> Through the fertile creation of this teaching, which under his influence and by his active participation, went from strength to strength, Charles Dupin must justifiably be considered, after Vaucanson, as the second founder of the *Conservatoire des Arts et Métiers*. (A. Morin 1873)

Jacques de Vaucanson (1709–1782) was a renowned inventor who is credited with the creation of the first true robots and the first completely automated loom. He had a fine collection of machines, which he left to the nation. They formed the basis of the conservatoire's collection.

Dupin's first lecture took place on 2 December 1820. His father felt obliged to congratulate him and did so on 11 January 1821 in the following terms:

> To M. Dupin, member of the *Institut royal de France*, rue Saint-André des Arcs, n° 33, Paris.
> Of all the arts and crafts taught at the *Conservatoire*, I know of none more interesting, my friend, than the art of letter-writing, or if you will, the craft of writing, when it is between people who are worth the trouble, for example from fathers to children. You wished me a happy new year. I wish you the same so, on that count, we are quits.

He said nothing more in this letter on the subject of the conservatoire—a topic which would have interested his son. Instead, Dupin père launched immediately into the subject of marriage and further dug at his son:

> I am not the one to press you, for I know too few women whose husband I should wish to be and too few children of whom I should wish to be the father. Meanwhile give your lectures, go on your errands and fill your purse. (DRB; PJ 77)

11.3. The Ensuing Years

Before examining Dupin's role at the conservatoire, I shall briefly summarize the years following his appointment to the conservatoire. In 1823, he was a member of the consultative council of the Arts et manufactures. Also in 1823, he became chevalier de l'ordre de Sant-Louis, a royal and military order under the auspices of the minister for the navy and the colonies; according to the ritual of this order, Dupin had to swear fidelity to the king and that he would never leave the service for that of a foreign prince. In 1824, he was made a baron at the same time as Thénard. The title had been abolished in 1790 and was restored by Napoleon I in 1808, but it could not be inherited. In 1824, Dupin became officier de la Légion d'honneur, a title he greatly desired.

As early as 1 September 1821, he had written to his elder brother asking him to use his influence in that direction:

> The Minister for the navy is well disposed towards me, but the minister is a man incapable of decisions; when the *Bureau* asked him to make me *officier de la Légion d'honneur*, he replied neither yes nor no. If you could write to Roy [then the minister of finance], that would be very useful to me. The 13th September, anniversary of the baptism of the Duke of Bordeaux, will undoubtedly be an occasion for new decorations.

The rest of the letter contains references to the works with which Dupin was engaged:

> The *Moniteur*, the *Constitutionnel* [the organ of Dupin's elder brother] and the *Courrier* reviewed the *Force navale* in lengthy and substantial articles. During the month of August, I already prepared much material for the third part of my *voyages* and many other things of particular use to our country, for that is the main aim of my work. (DRB; PJ 78)

This was a reference to the *Force commerciale*, the final part of the *Voyages dans la Grande-Bretagne*, which would be published in 1824.

To improve his chances of being appointed to the Légion d'honneur, even though the 13 September deadline for the award had passed, Dupin wrote from London on 9 October to the minister for the navy and the colonies seeking a promotion to officier. He wrote, 'For the seven years I have been a simple *légionnaire*, I feel I have earned this noble decoration' (ASHM). In fact, it had been eight years since 'member of the Legion of Honour' had appeared on the title page of his 1813 *Développements de géométrie*. In spite of all his efforts, he was not to be successful in this ambition until 19 August 1824.

This delay can probably be explained by the fact that he had several times upset the government. For instance, in the biographical index of the *Dictionnaire des parlementaires français* is this entry:

> His work *Force militaire de la Grande-Bretagne* (first of the three parts of the *Voyages dans la Grande-Bretagne*) angered the government because of the liberal tendencies of the author; the work was referred by the Minister for the navy [who, as has been seen, was unable to make decisions] to the Council of ministers and censored; Ch. Dupin, having refused to submit to these decisions, was for four years in disgrace. Finally, the government relented and made him *officier de la Légion d'honneur*. (Robert, Bourloton, and Cougny 1891, 493)

It has also been seen in chapter 10.4 that the 1823 publication of the *Système de l'administration britannique, en 1822* had aroused Villèle's animosity.

However, the future now looked brighter, and on 27 July 1832, Dupin would be promoted to commander and then to grand officer on 27 April 1840. In 1826, he requested permission from the Royal Order of the Legion of Honour to accept the title of Knight of the Order of the Sword, conferred on him by the Swedish king. In 1827, he was to be elected député and in 1832 a member of the Académie des sciences morales et politiques. This was the very year of its reestablishment as an academy to replace the Institut's *classe des sciences morales et politiques* (moral and political sciences section), which had been abolished by Bonaparte in 1803. Dupin joined the political economy section and was to present several papers to this academy, including, in 1849, his *Situation actuelle de l'industrie française* (Current situation of French industry).

This has been a summary of Dupin's ascent in the world. I shall now return to the changes in his career that resulted from his visits to Britain.

11.4. DUPIN AS AN EDUCATOR

Beginning in 1818, Dupin was a resident in Paris. His appointment as a professor at the Conservatoire des arts et métiers marked a change of direction in his career and an end to his creative work in mathematics and mechanics. It is true that his *Applications de géométrie* appeared in 1822, but it grouped together his previous works. In the Génie maritime, his rank as a superior officer took him away from engineering.

Instead, he turned to the role of an economist and an educator. He was a professor, but he was more than a teacher; not only did he teach geometry and applied mechanics, but he also aimed to prepare his students in the moral sphere and to ready them for professional life. He wanted to direct them towards—and this was his great concern—the bettering of the conditions of the workers. He was, in all ways, an educator—an *éducateur* in the French sense of *éducation* as meaning 'good breeding'. He would pursue his aim of educating the workers, not only at the conservatoire and in his other teaching, but also when he created, in the main military or commercial ports and in various towns in the interior, courses in the style of those at the conservatoire or established commercial schools for young people from modest backgrounds. This concern would continue even in his various functions apart from teaching—in the sphere of economics, in the creation of savings banks, and as a politician—in spite of the accusations of paternalism or demagogy which were raised against him in certain quarters.

Living, in 1818, in the rue Saint-André des Arcs, he would cross the Pont-au-Change and walk down the rue Saint-Martin and across the market to enter the conservatoire by the great gate of the white stone and red brick building which today forms the centre part of the vast establishment. What was the conservatoire? I shall study that institution in chapter 12, together with the role that Charles Dupin played in it.

ENDNOTES

1. General J. V. Poncelet (1788–1867). See Chatzis 2009 for more on Poncelet's work.
2. A. Lavallée (1791–1871), the founder of the Ecole centrale des arts et manufactures in 1829 as a private initiative, with the chemist and politician J.-B. Dumas (1800–1884), the physicist E. Péclet (1793–1857), and the mathematician T. Olivier (1793–1853) as its first professors. It was a private school aiming to produce consultants and experts for industry and factories. For the history of the school, see Pothier (1887).

Chapter 12

The Conservatoire national des arts et métiers

12.1. France at the Time of the Bourbon Restoration

The French monarchy, in the person of Louis XVIII, was briefly restored in 1814, but this was followed by Napoleon's return and the Hundred Days of his second reign. After the emperor's final downfall, Louis XVIII again returned to the capital in July 1815, with a subsequent period of the White Terror and ultraroyalist reaction. France found itself in a difficult political and economic situation. Roads and bridges had been neglected or damaged, the merchant navy was in disarray, and France was suffering from the crippling financial burden imposed on it by the treaty following the Hundred Days. The country was having to pay vast sums in compensation to its neighbours and support an occupying force. Scientific discovery and development in the interests of industry were recognized as essential. However, there was scant concern for the education of the lower classes, and the monarch retained the imperial organization. The government was imposing strict ideological control on educational institutions, the Ecole polytechnique was closed, and Gaspard Monge, one of its founders, was categorically dismissed from all his duties. The church was playing an increasingly important role in education, and in 1827, this was to culminate in the Education Ministry's becoming the Ministry of Ecclesiastical Affairs and Public Instruction. The repressive atmosphere and the absence of career prospects meant that highly trained engineers, for example, were seeking work abroad (see Bradley 1981).

On the positive side, it was now possible for people to cross the Channel, and many observers did so, Charles Dupin among them, to observe British industrial successes. France had to make herself competitive, and it was not enough for her to develop industry; the manpower to run it was also a prime necessity. This involved educating the workforce in the management and understanding of new machinery and the organization of work. Dupin, who had been so greatly impressed by the work of Ure (see chapter 10.3), was to play a leading part in emulating that work in France at the Conservatoire des arts et métiers.

12.2. A Brief Survey of the History of Technical Education in France

Before looking at Dupin's new career at the conservatoire, I shall briefly examine the situation of technical education in France at that time. The first big technical school was probably the Académie royale d'architecture, founded in 1671, where there was teaching of practical mechanics, stonecutting, and civil engineering. Soon afterwards, in 1682, Jean-Baptiste Colbert established the Ecoles nationales de navigation (National schools of navigation), which taught mathematics, drawing, and cartography, but it would seem that future naval officers showed scant interest in the technical side of their profession.

Throughout the seventeenth century and even later, elementary technical education was provided mainly by private establishments, lay or religious. An example of this was the school at Saint-Yon, founded by the Abbé de la Salle in 1680 for children of the poor, who received teaching in mathematics and architecture but not in Latin.

The eighteenth century saw the creation of several engineering high schools—for example, the Ecole des ponts et chaussées (founded in 1747); the Ecole du corps royal du génie at Mézières (founded in 1748), where Monge taught descriptive geometry; the Ecole des ingénieurs constructeurs de vaisseaux (founded in 1765); and the Ecole des mines (founded in 1783). At the end of the century, these schools were dominated by the Ecole centrale des travaux publics, which was created in 1794 and a year later became the Ecole polytechnique (see chapter 3.2). The other schools were then application schools for Polytechnique graduates. The nineteenth century produced the Ecoles d'arts et métiers, the Ecole des mines at Saint-Etienne, courses at the Conservatoire des arts et métiers, the Ecole supérieure de commerce in Paris, the Ecole des hautes éudes commerciales (HEC), and the écoles nationales professionnelles (ENP). The level of teaching in these establishments had been raised to a considerable degree since their creation, but they produced only *sous-ingénieurs*, senior technicians, or overseers. That, indeed, was the justification for the private initiative that created in 1829 the Ecole centrale des arts et manufactures, which was to produce qualified engineers for industry. In the nineteenth century, engineers were educated at the Polytechnique, the Ecole normale, and the university faculties. Comberousse commented,

> For certain minds, seeking after truth is sufficient; it becomes a noble passion to which all interests are sacrificed. For the mass of intellects, on the other hand, the study of the sciences is but a useful adjunct to learning. It is therefore necessary to provide them with clear, quick teaching, full of fact and rich in precise rules.
>
> It was in this category that the *Ecoles d'arts et métiers*, the Saint-Etienne *Ecole des mines* (at least at its beginning) *and the Ecoles nationales professionnelles* may be placed. (Comberousse, 1879, 8–9)

He continued,

> The schools in Châlons, Angers, Saint-Etienne … prepare overseers only for certain specialisms. Finally, between these two categories, there is a third, no less important. It comprises all men whose career obliges them constantly to apply scientific concepts. For them, science is, in general, neither a passion, nor a simple increase in culture; it becomes an indispensable tool … It is for this third category of minds that the *Ecole centrale [des arts et manufactures]* is intended.

12.3. The Conservatoire des arts et métiers

Created in the same year as the Ecole centrale des travaux publics (subsequently the Ecole polytechnique), the aim of the Conservatoire des arts et métiers was not to produce state engineers but rather to foster technical innovation among the skilled workers and industrialists of Paris. It was the brainchild of Henri Grégoire (1750–1831), a former priest and bishop of the Loir-et Cher, then a député in the Convention and a reporter for the Commission of Agriculture and the Arts. It was he who presented to this assembly the plan for the creation of the new institution, which was ratified by a decree of 18 *vendémiaire* (10 October 1794) (see Payen 1998).

The terms of the decree were:

> There will be formed in Paris, under the title of *Conservatoire des arts et métiers*, and subject to the inspection of the Commission of agriculture and the arts, a public repository of machines, tools, models, designs and books of all types related to the arts and crafts. The originals of invented machines and instruments will be lodged at the *Conservatoire*.
>
> There will be explanations of the construction and the use of tools and machines useful in the arts and crafts.
>
> The *Conservatoire des arts et métiers* will comprise three demonstrators and a designer. (Payen 1998)

According to Dulaure (1821, 28), three repositories of machines formed the basis of the conservatoire—one at the Louvre, another at the Hôtel de Montagne (rue de Charonne) containing the machines left in 1782 by Vaucanson, and the third in the rue de l'Université. Moreover, it was thanks to Grégoire that the Conseil des Cinq-Cents allocated a large part of the buildings of the ancient abbey of Saint-Martin-des Champs to the conservatoire in 1798.

In addition to Grégoire, Charles Alquier of the Conseil des Anciens played a part in the evolution of the institution. Its final home in the ancient abbey was effected in 1799; the former refectory of the monks became its library. Confiscations of possessions and goods during the revolutionary period provided much of the material for the new schools, and the full warehouses were a rich source for the Ecole polytechnique, for example, as well as for the conservatoire (see Oliver 2005).

Dulaure noted that various changes occurred in the administration of the conservatoire. In 1810, a free school was founded there to train young people to become skilled and educated craftsmen and distinguished professors. In this school, the subjects taught included life drawing, ornamentation, architecture, algebra, geometry,[1] and the application of these various branches of mathematics to woodcarving, stonecutting, and the calculation of machines. Nearly a decade later, great changes took place, as Dulaure described: 'In 1819, three chairs were created, one in industrial economy and the two others in chemistry and industrial mechanics applied to the crafts' (1821).

It should be added that in 1810, a school of cotton spinning was established at the conservatoire, which during the Empire, along with the drawing school, would constitute the main area of teaching within the establishment. It was entrusted to Jacques Milne, a member of a famous family of cotton industrialists of English origin. However, it fell victim to the 1814 economic crisis and disappeared.

At first, the free school had no teachers, instead relying on demonstrators who worked the machines and explained their mechanisms. There was a machine construction workshop for repair work, for creating new machines, and to put rare and expensive tools at the disposal of builders. There was also a drawing office where plans and mechanisms could be consulted and copied. The students ranged in age from twelve to nineteen and sat for an end-of-year examination that resulted in a distribution of prizes (technical books) presided over by Hachette, Dupin's professor of mathematics at the Ecole polytechnique. The library contained French and foreign books on the various arts and sciences (Grelon 1996, 46). The conservatoire was a landmark in the creation of French technical education.

At the end of the Empire, the teaching of geometry and design ended for a time but was restored in October 1816 by the new government. The organization remained much the same, except that the minimum age of entry was raised to fourteen. However, the teaching was to undergo fundamental changes inspired by Nicolas Leblanc, the son of the man who invented the process of fabricating soda from sea salt. Henceforth, drawing would no longer be an end in itself but would be used

> to facilitate the representation of machines, to popularize, as it were, the understanding of their functions and the interconnection of their different parts … not only draw, with perfect accuracy the machines already functioning, but also to produce, in advance, designs for the construction of all kinds of combinations or apparati still in the realm of invention. (Leblanc 1830)

Leblanc took the students on weekly visits to workshops in the region, and they also attended lectures by the physicist Jacques Charles[2] on the physics and analysis of machines. This emphasis on application was intended to meet the needs of industrialists and machine builders in the Paris region and resulted in an ever-increasing number of students.

In 1819, there was to be another great change. As Grelon put it,

> This initiative was largely due to the efforts of Baron Charles Dupin, polytechnician and engineer in the *Génie maritime*, who had made numerous visits, starting in 1816, to Great Britain, producing an account of his journeys in six large volumes. Dupin, like many scientists, was convinced that new teaching was necessary in burgeoning disciplines like mechanics and chemistry, aimed at industrialists, technicians and workers, to develop innovation and to catch up with England in her state of

> prodigious industrial superiority … Dupin had seen in Glasgow an institution that had greatly impressed him, described at length in his *Mémoires sur la marine* (1818). This was the Andersonian Institution offering evening classes to young workers at minimal cost. For Dupin, it was clear that the industry and commerce of the town had greatly benefited from the results of this training. (1996, 48–49)

12.4. Dupin Opens His Conservatoire Courses

Dupin arrived at the conservatoire imbued with a sense of a social mission and with the firm intention of using the lessons he had learned in Scotland, which were still fresh in his memory. As he had written in 1817 in the report of his second visit to Britain,

> The wealthy inhabitants of Glasgow have founded the Andersonian Institution, where during every winter evening, are taught the elements of geometry, mechanics, physics and chemistry applied to the arts. These courses are aimed particularly at young skilled workers. The Andersonian Institution has produced astonishing results. It is a wonderful thing to see today, in many of Glasgow's workshops, simple workers with a knowledge of, and an ability to develop when necessary, the principles of their operations and the theoretical means to arrive at the most practical and perfect possible results. (1818b)

Dupin was to teach Méchanique [*sic*] appliquée aux arts (applied mechanics), and his lectures would be aimed at an audience with little basic knowledge. He was well aware of this, and indeed, it was what he had wanted. However, he had also given thought to the usefulness of lessons in practical mechanics for educated people, whose knowledge was purely abstract (see Monzie 1848). This is illustrated in three letters, the first of which was written by Dupin to the minister for the navy and the colonies on 9 October 1820:

> a. I shall shortly begin my lessons in *méchanique appliquée aux arts*, as decreed by His Majesty.

b. My dearest wish: that this course be useful to the arts of the navy.
c. So, this year, the *Polytechnique* will provide several students for the *Corps du génie maritime*.
d. I therefore propose that those young men spend the winter in Paris, in order to attend my lectures, before they are sent away to the ports. (ASHM CC7 779; PJ 79)

The second letter is dated 23 October 1820 and was addressed by Pierre-Jacques-Nicolas Rolland, the inspector general of the Génie maritime, to the minister for the navy and the colonies. It makes the following points:

a. M. Dupin, naval engineer, professor of *mécanique* at the *Conservatoire des arts et métiers* has proposed that the *Polytechnique* graduates proceeding to the *Génie maritime* should attend his lectures.
b. Here are my observations:
If these students stayed for a year in Paris, these courses would be their sole occupation.
c. So, the families of these young men could fear that their children, left to themselves,[3] at a susceptible age, might fall into wicked ways.
d. This risk would not to be feared in the ports, for the opportunities to indulge in vice are fewer than in Paris. (ASHM CC7 779; PJ 80)

Dupin was not impressed by these arguments, and he again wrote to the minister for the navy on 4 November 1820, making the following statements:

a. The *Polytechnique* students joining the *Génie maritime* must receive practical teaching in addition to their technical education. Thence, the usefulness of following my lectures at the *Conservatoire*.

b. Young people certainly do not have fewer temptations in the ports than in Paris, for gaming, wine and women. (ASHM; PJ 81)

As it was, with an income of 1,200 francs a year, it would have been difficult for the students to indulge their fancies. In any case, Dupin was not successful in this plan. It might have been disastrous to put Polytechnique graduates in the same class as comparative beginners.

12.5. Dupin's New Colleagues

Two colleagues were appointed to join Dupin at the conservatoire, Nicolas Clément-Desormes and Jean-Baptiste Say. Clément-Desormes (1779–1842) was a physicist and an author whose particular interest was the measurement of the relationship of specific heats of gases. Say, who is much better known today than the physicist is, was the author of *Traité d'économie politique* (A treatise on political economics; 1803). He may be considered as having popularized the economic theories of Adam Smith, the author of *An Inquiry into the Nature and Causes of the Wealth of Nations* (a work which, incidentally, had also inspired Prony during his time at the Cadastre [see Bradley 1998]).

These three men—Dupin, Say, and Clément-Desormes—would be responsible for the teaching at the conservatoire as a veritable teaching establishment, with their respective chairs in mécanique appliquée aux arts, industrial economics, and chemistry applied to the arts. According to a report in the *Annales de l'industrie nationale et étrangères*, when their courses began,

> A studious audience crowded into this establishment where they were to hear, successively and on the same day, four men whose services and talents are dear to industry and the sciences. M. le duc de la Rochefoucauld, who treats all subjects with his customary eloquence, especially when they are related to the industrial arts, opened the session as chairman of the administrative

> council and his speech was loudly acclaimed. M. Dupin came next and, as professor of mechanics, he outlined the plan he would follow in his courses. Next came M. Clément who, in a most instructive talk, developed the series of improvements that the industrial arts had achieved and could still achieve through the study of chemistry. M. Say (J-B.), appointed to the course in industrial economics, shed the most advantageous light on this course, and inspired all his listeners with the desire to attend it.
>
> The unanimous applause must have proved to the learned professors that they could easily satisfy an educated and severe audience. (Ragey 1970, 23)

As for the prevailing atmosphere during the lectures when they began, a police report provides the details because Charles X, at the beginning of his reign, imposed strict surveillance over the conservatoire. With regard to Dupin's lectures, the report states,

> These meetings are numerous and the hall in which they take place is equipped as an amphitheatre … many young people from manufacturing industry may be seen there, some mathematics students, young people from the *Ecole spéciale de commerce* … the deepest silence reigns throughout the lectures. (Ragey 1970, 24)

The comparatively liberal Decazes government under Louis XVIII did not last long. As Grelon put it,

> After a brief period of transition, it was an ultraroyalist government led by Villèle, which took power. Thus the *Conservatoire* experience would be more or less sabotaged by the powers that be, fearing the teachings of Dupin and Say: the police, as a result of reports from informers, reproached them with criticising government action during their lectures. In order to limit the audience and change its social composition, the evening classes were forbidden and there would be no more scholarship students. All seemed to be aimed at the disorganization of the institution. It was only after Villèle's ministry ended, in 1828, that evening classes would be re-established. However the initiative of the *Conservatoire* would

> not be lost. The concept of integrated industrial education was on course. (1996, 49)

Grelon added that Dupin managed to continue to develop industrial education in the provinces. On 25 December 1825, the minister of the interior sent a circular to *préfets* (local administrators) asking them to support Baron Dupin's initiatives and to institute courses in mechanics in the main towns of France.

12.6. Dupin's Opening Lectures

With his appointment as a professor of mechanics applied to the arts, Dupin had the idea of launching his courses annually with an opening lecture on a subject of general interest in the fields of science or technology, a social issue, or education (see Perrin 1983, 440–459). He gave several general lectures at the conservatoire and also at the Institut de France and the Chambre des députés (see Grattan-Guinness 2008, 26).

His output was prolific; for the period of 1819 to 1840, Grattan-Guinness (1990, 1106–1110, 1463–1465) has cited more than fifty publications and eighty different publishing locations for Dupin,

> either as a small book produced by Bachelier and/or in a newspaper like the *Moniteur universel* and/or in a journal like the *Revue encyclopédique*, and/or in a technological journal as, for example, two series in Ferrusac's *Bulletin*, the *Recueil industriel* and the *Annales de l'industrie nationale*; the majority of these were the texts of his lectures; the others are either essays on subjects closely related to an opening address or concerning one of his lectures.

In addition, there were two volumes of reeditions, other speeches, and other information (Dupin 1825a); his lessons in geometry and mechanics at the conservatoire, published in three volumes (1825b, 1825c, 1826b); two volumes of descriptions and statistical analyses of all the French départements (1827a); and six short volumes on manufacturing, written for workers and sold at an affordable price (1827b). For more on Dupin's publications, see the bibliography to this work and Vatin (2007).

Grattan-Guinness observed that much of the content of Dupin's lectures is still relevant—for instance, his work on the use of domestic animals (given at the Athénée in 1827) or on the world's population and its resources (given at the conservatoire in 1830). However, he concentrated mainly on four of Dupin's opening lectures. On those of 1824 and 1825, he stated,

> Dupin, then Professor of mechanics and geometry applied to the arts, considered it necessary to improve the level of technical and scientific knowledge amongst the working classes. He suggested that highly educated savants, *polytechniciens* in particular, should offer in their towns and regions free courses in geometry and mechanics, on the pattern of his own work at the *Conservatoire*. He made several other speeches on this topic. In 1828, during a tour of the main French ports, he himself gave the opening lecture for the new courses created in Bordeaux and Toulouse. His speeches were on very varied themes. For example, in 1826, he wrote a special report for the Minister for the navy on the advantages of introducing such courses in the naval towns; this document was published at least eight times during that year, in either a complete or an abridged version.
>
> This pedagogical project was well received: a number of his colleagues followed his initiative. The case of Metz was the most remarkable. (2008, 27)

It was Claude-Lucien Bergery (1787–1863) who, with General Jean-Victor Poncelet (1788–1867), proposed a free course for young workers in Metz, to be run at the expense of the Sociéte des lettres, sciences et arts et d'agriculture de Metz. Inspired by the work of Dupin, Poncelet, who was already a professor of applied mechanics at the Metz Ecole d'artillerie et du génie, began this new venture in 1827 (see Vatin 2007 and Chatzis 2009). There was a certain ethical spirit among the teachers who participated in this movement. *Polytechniciens* who had been privileged with an excellent education at the Ecole polytechnique and the application schools, such as the Ecole des mines or the Ecole des ponts et chaussées, had achieved great success in life as a result of their training. They felt it necessary to show their gratitude for this and to give something back to France. This attitude was somewhat paternalistic and was not always appreciated, as Jarrige showed in his '"Nous ne sommes ni vos élèves, ni vos amis": Dupin et les ouvriers face à la presse ouvrière sous la monarchie de juillet' (We are neither your pupils nor your friends: Dupin and the workers before the workers' press under the July monarchy; 2009).

Grattan-Guinness (2008, 27) pointed out that in the third volume of his own courses, Dupin provided a general overview of ergonomics, to use the modern term. He analyzed the capacity and the limits of manpower, both intellectual and physical, as he raised questions of fatigue as well as the use of animals (see Dupin 1826b, lessons 4 and 5). This was a period of some innovation in technology, but Dupin's efforts were to be eclipsed to a degree by the establishment of the Association polytechnique, a society whose aim was providing free and public education to workers. It was founded in 1830—at the end of July, another turbulent time in French history—by Polytechnique students and was designed to make workers more skilful and better equipped to cope with machinery.

In that year, Charles X decided on a coup d'état, dissolved the Chamber of Deputies, and curbed the freedom of the press. Armed resistance came from the republicans, supported by workers, former officers, and *polytechniciens*; barricades were erected and three days of bloodshed began. Charles X panicked, but the republicans were aware of how few they were in number. However, Charles fled to England, and the result was the establishment of the July, or bourgeois, monarchy in the person of Louis-Philippe. So, France broke again with the Ancien Régime, took up the national colours, and determined that never again would there be a return to privilege. The Association polytechnique, with Auguste Comte as one of its propagators, was therefore born after a fight for freedom in which *polytechniciens* and workers fought side by side in the defence of their rights. They were to join each other again as tutors and students in the association's classes. Starting in 1830, there were fourteen free lectures given every week, and the number continued to increase until 1848. Vatin (2007) pointed out that the association was based on the model of the one at Metz, which also preceded Charles Dupin's movement.

At this point, another school should be mentioned, the Paris Ecole centrale des arts et manufactures, which was founded in 1829 with the aim of providing higher education for workers in industry and commerce. (See Grattan-Guinness 1990, 1113). This was a private enterprise that was financed by Alphonse Lavallée, a young landowner, and founded by three scientists, the chemist Jean-Baptiste Dumas, the physicist Eugène Péclet, and the mathematician Théodore Olivier. The minister of education, Antoine-François-Henri Lefebvre de Vatismesnil, strongly supported the project. The aim of the school was to provide graduates in industrial engineering, for which the Ecole polytechnique and the *écoles d'application* (application schools) had not been designed. With regard to this school, there is an indication of extraordinary ignorance on Dupin's part, because in 1830, he proposed the inauguration of such an institution to the Chambre des députés. The founders of the already existing school were obliged to write to him to announce their existence (Pothier 1887, 69–70, 455–456).

12.7. Themes for the Opening Lectures at the Conservatoire

In 1826, Dupin chose as the subject for his opening lecture the development of primary education. He had calculated simple statistics for each département: the number of places that were available in schools for male children divided by the total population. He explained, 'To make the main difference apparent, I had the idea of giving the various departments an increasingly deeper colour, when they were sending fewer male children to primary school.' (Dupin 1827a, 2:249)

The map, and the details of its preparation, have been described by Palsky (1990, 1996). It was probably unnecessarily large in scale, but the very black zones, illustrating the worst areas for schooling, stand out, illustrating Dupin's interest in the visualization of learning. The map was sold separately in a reduced format, and it became very famous. Daniel Colladon, a Swiss engineer who was living at that time in Paris, raised the question of the map during a visit to the Lot-et-Garonne and mentioned his personal acquaintanceship with Dupin (Colladon 1893, 162–163). The Lot-et-Garonne department was grey rather than black.

It should be mentioned that Dupin was not the only member of his family to take an interest in statistics. His cousin, Claude Dupin (1767–1828), the prefect of the Deux-Sèvres department from 1800 to 1813 (who married, in 1800, the widow of Georges Jacques Danton, one of the leading figures of the French Revolution, guillotined in 1794), had worked in the subject much earlier. In 1804, he produced his *Mémoire statistique du Département des Deux-Sèvres* (Statistics of the Deux-Sèvres department), a comprehensive survey of the department and its inhabitants.[4] This may well have influenced Charles Dupin in his work.

Dupin, now a professor of geometry applied to the arts and statistics, celebrated the twentieth anniversary of his courses by giving a lecture on the history of the teaching at the conservatoire—not only his own, but also that of his two colleagues and fellow founders, Jean-Baptiste Say and Nicolas Clément-Désormes (Dupin 1839b). As to the future of the teaching in general at the conservatoire, he expressed himself thus:

> Rational or theoretical geometry has as its aim the definition of forms, the measurement and drawing of dimensions, in fact to show the relationships between expanses, either of space or the bodies occupying part of that space: its truths are absolute, its laws are general and its relationships completely rigorous. On the other hand, the geometry of the arts is a geometry of approximation. It measures surface distances, volume; but with our sense of observation, our hands, our tools, our instruments, we can achieve only an approximation of the rigorous measurement … If the industry of a people is in its infancy, the gross approximation is far removed from the rigorous ideal. But as this people perfects its industry …, so also the products of its skills are less

> far removed from the precision which geometry and mechanics define and demonstrate. (1839b)

This lecture was greeted by deep silence, apart from occasional lively expressions of appreciation, but as Dupin came to the end of his presentation, there was thunderous applause and repeated demands that the speech be printed. Dupin was a fine, if a loquacious, orator.

FIG. 1. Charles Dupin's 1826 map of the state of education in France.

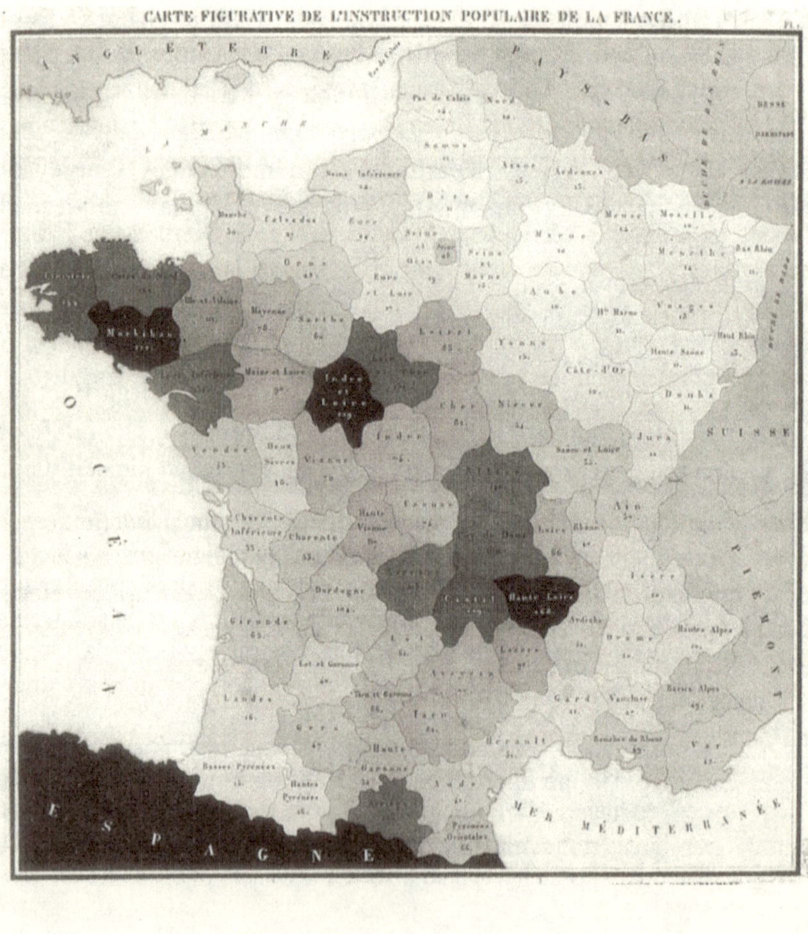

Endnotes

1. Incorrect for 1810. Geometry was not yet taught there.
2. Jacques Charles (1777–1859), physicist and engineer, known for his work in acoustics, heat, and pressure.
3. The students graduating from the Polytechnique were sent back to their families on vacation for two months before joining the corps.
4. I am grateful to Pierre Ménard for drawing my attention to this family connection.

CHAPTER 13

GEOMETRY AND TEACHING

13.1. THE INAUGURATION OF THE AMPHITHEATRE AT THE CONSERVATOIRE IN 1822

Upon the occasion of the inauguration of the amphitheatre at the conservatoire in 1822, Dupin spoke once again about the importance of encouraging French industry. He praised the work of the minister Decazes, who had established councils for commerce, industry, and agriculture in 1819; facilitated the fifth national industrial exhibition; and as has been seen, established three chairs of applied science at the conservatoire. Dupin then launched into a vehement attack on the corporations (see Mertens 2004, 153) which had held industrial development in a tight grip and had controlled the training and the privileges of trade. He claimed that there had once been some justification for them, 'to defend themselves against the extortions, acts of injustice and the demands of the powerful idle class' (1822g).

Hostility to the French corporations had peaked during the revolutionary period, and Dupin claimed that at the beginning of the seventeenth century, they had already lost their raison d'être because royal power had curtailed the arbitrary despotism of local lords. Turgot[1] abolished the corporations in 1776, but they were subsequently reestablished. It was not until 1791 that they were finally abolished by the National Assembly, and although that saw an end to their negative practices, it also affected the ancient system of apprenticeship. An alternative system of technical education had to be provided. Dupin ended his speech with references to the écoles d'arts et métiers of Châlons-sur-Marne and Angers as well as to his own course in applied mechanics at the conservatoire. It was during this year that a work he considered to be of vital importance to his teaching appeared, the *Applications de géométrie et de méchanique* (Applications of geometry and mechanics).

13.2. THE *APPLICATIONS DE GÉOMÉTRIE ET DE MÉCHANIQUE*

Among the works that were drafted in England by Charles Dupin was his *Applications de géométrie et de mechanique* [*sic*]. It is now time to look at it, as it has been seen that applied geometry was a great focus of his courses at the conservatoire, and this work was essential to his teaching there. It appeared in 1822, while Dupin was evolving from an engineer to an economist and an educator. In essence, it is a grouping, with some modifications, of five already known papers, and it adds little to what Dupin had written in 1814 and 1816. Rather than a creative scientific work, it is a didactic collection and is therefore relevant to the present chapter. As Dupin put it in the 'Considérations préliminaires', 'I now bring together, in one volume, the applications I have made of geometry to various questions concerning public works' (1822a).

The work responded to a wish expressed by Carnot, Monge, and Poisson, who had written in their report of the first three papers of the *Développements de géométrie* (in December 1812; see chapter 3.6), 'The research that we have discussed elicits the desire that an engineer with such extensive knowledge of geometry and analysis, soon publish the work in which he proposes to apply it to practical questions of public usefulness' (Procès-verbaux de l'Académie de sciences, 14 December 1812). That was what Dupin had finally done after nine years.

The *Développements de géométrie* had first appeared in 1813 (see chapter 3.6). Now, the contents of the *Applications* were as follows: First was *De la stabilité des corps flottans* (The stability of floating objects), a paper that had originally been presented to the first class of the Institut de France on 10 January 1814 and was deemed worthy by the reporters Sané, Prony, and Carnot of insertion in the Institut's collection of the *savans étrangers* (external scholars). It was an account of the first practical application of the methods that had been published in the *Développements de géométrie*. The second paper, *Du tracé des routes isolées* (on plotting isolated roads), had been presented to the first class of the Institut on 25 September 1815. The third, *Sur le tracé des routes, dans les déblais et les remblais* (on plotting routes through cuttings and embankments), was based on Dupin's (circa) 1804 *Mémoire sur les déblais et les remblais*, with additions from his personal experience, and had been presented to the Institut on 18 December 1815.

Paper number four, *Sur les routes suivies par la lumière et par les corps élastiques en général, dans les phénomènes de la réflexion et de la réfraction* (the course followed by light and elastic bodies in general, in the phenomena of reflection and refraction), was presented to the Academy of Sciences on 22 January 1816. In 1816—and still in 1822, the date of publication of the *Applications*—the notion of optics was very much in contention between the emission hypothesis of Newton and Huygens's theory of undulations. The fifth paper, *Examen de la structure des vaisseaux anglais* (the structure of English ships), was an augmented version of the one that had been read to the Royal Society of London on 19 December 1816 and published in the *Philosophical Transactions* in 1817 (see chapter 7.4).

From this compiled work—the *Applications de géométrie et de méchanique*—it is apparent that where questions of physics were concerned, Dupin was essentially a mathematician and an engineer rather than a physicist. Indeed, he had said as much himself (see chapter 4.3

and PJ 17): 'You will see that I have reduced optics to being but a particular case of the theory of cuttings and embankments'.

Cuttings and embankments were constant preoccupations of engineers at this time. It was this work for which the young Malus[2] had been noticed by a *Ponts et chaussées* engineer, as he had found a means of cutting and banking as easily as possible for military engineers; this gained him entry to the Ecole polytechnique (Chappert 1977). The theory of cuttings and embankments to which Dupin brought optics was, in its turn, attached to the theory of curvature. Dupin took gentle issue with Malus, who had thought that his results were valid for only a single reflection or a single refraction. Dupin commented,

> As Malus used an extremely complicated analysis, a single error in calculation could have convinced him that it is no longer possible for reflections on a second mirror to satisfy the equations of condition on which he depended the orthogonality of the developable surfaces formed by the reflected rays. But the erroneous consequence arising from this somewhat mechanical error, in no way detracts from the merit of having discovered one of the finest theorems of geometry applied to optics. (1822a)

Dupin's theory of cutting and embanking, to which he brought optics, connected with his theory of curvature, as he had declared in his letter of 1806 to Mathieu (see chapter 4.3 and PJ 15). He was, above all, a geometer specializing in the notion of curvature, and in this, he was very much the spiritual son of Monge.

13.3. STENDHAL'S COMMENTS ON THE *APPLICATIONS*

The renowned French author and politician H. B. Stendhal (1783–1842), whose comments on Dupin's choropleth map have already been seen, took a great interest in the exact sciences, perhaps rather for the discipline they imposed than for the sciences themselves. (The choropleth map will be explained in chapter 14.5). He had himself envisaged having a scientific career, and in 1799, he had gone to Paris intending to sit for the Ecole polytechnique entrance examination. He soon changed his mind, and the fact of his intention is hardly convincing of his ability as a scientific critic. However, that is the role that he chose to assume on a single occasion and somewhat superficially in his account of Charles Dupin's *Applications de géométrie*. It appeared in 1822, the same year as Dupin's work.

Stendhal had little time for French savants:

> All our savants are frivolous, and think only of paying court and going to church; very few work with a good conscience. The public recognizes exceptions in MM. Arago, Boissonnade, Courier[3] and Dupin.
>
> In M. Dupin's latest work we recognize the worthy pupil of Monge. The basis of the mathematical knowledge of that great man was a fine logic. He had begun life as a stone-cutter in Metz and it was from there that he went to rise, by merit alone, to the rank of senator, after having been one of the founders of the *Ecole polytechnique*. M. Dupin, well-known in England, is one of the finest students of that school which has just been disorganized in 1822, and which has given 4,500 distinguished subjects to France. His 'considerations' are an excellent supplement to all elementary courses in geometry. (Mérimée 1855, 210)

Stendhal's comments also appeared in England in the *New Monthly Magazine and Literary Journal* (1822, vol. 4). This version differs slightly from the French one:

> It is a rare occurrence at present to see a really clever work appear from the 'savans' of the institute. They make more profit these days from paying court to people in high places and being regularly seen at mass, the observation of the external rituals of religion now being essential. To this display of self-satisfaction, there are certain honourable exceptions, of whom M. Dupin is one. In this work he is revealed as a worthy pupil of Monge, for whom the basis of mathematical knowledge was sound and infallible logic. That extraordinary man began life as a stonecutter in Metz and it was from that modest position that he rose by his own

> merit and without charlatanism to the rank of senator. He may be considered as the founder of the *Ecole polytechnique*. M. Dupin, well known in England, is assuredly one of the best products of that establishment which has given so many remarkable men to France and which, extraordinary thing, has just been suppressed. These '*considérations générales*' are an excellent addition to all the usual elementary courses in geometry.

Stendhal's comments hardly constitute deep scientific criticism, but they have a certain interest, emanating as they do from one of France's greatest literary authors. As regards the suppression of the Ecole polytechnique in 1822, the school's council for improvement had long been trying to tackle the frequent disorder and political activity among the students. Constant vigilance, expulsions, the separation of the two divisions of the students so that they could not meet—all of these attempts at maintaining order had failed. In the new repressive Bourbon climate, it was decided to return to the severe discipline of a military regime. The school's governor could now, in the event of a report from the deputy governor, suspend staff, professors, tutors, and employees and provisionally suspend students from the school.

13.4. Dupin's Work at the Conservatoire Continues

When Dupin's *Applications* appeared, he had been teaching for two years at the Conservatoire des arts et métiers. Inevitably, the work went far beyond the level of his students, and it was not until 1825 that his first pedagogical work, *Géométrie et mécanique des arts et métiers et des beaux-arts* (1825b), was published for them. It corresponded to his teaching and was aimed at craftsmen, workers, overseers, and workshop managers. His four years at the conservatoire enabled him to produce this work.

There were four periods in Dupin's work at the conservatoire in addition to the two that have already been indicated. These were 1820 to 1824; 1825 to 1839; 1839 to 1852, with a course entitled 'Geometry Applied to the Arts and Statistics'; and 1852 to 1864, with a course entitled 'Geometry Applied to the Arts'. In 1864, Dupin was very old and was often replaced by his assistant Aimé Laussedat of the Academy of Sciences. When Dupin died in 1873, Laussedat earned the title of professor.

In his *Discours et leçons sur l'industrie, le commerce, la marine et sur les sciences appliqués aux arts*, the influences of Dupin's travels in Britain and his vocation as an educator appear very clearly:

> To the workers of France.
>
> My friends, I dedicate to you the work that I have had the greatest pleasure in writing. I went to the land of our rivals in industry; I saw how savants and men of power united their efforts to provide the English, Scottish and Irish workers with new education, to make them more competent and wiser. I wanted for you the same privilege and even more, I thought we could give you more complete and more advantageous teaching. (1825c)

Dupin quoted the pioneering work of Benjamin Franklin (1706–1790), the successful contribution of Richard Arkwright (1732–1792) to the textile industry, the work of James Watt (1736–1819), and the work of the scientist and encyclopaedist Jean le Rond d'Alembert (1717–1783). Then, he added,

> Workers, those are examples that are meaningful; they will suffice to inspire with a noble zeal those of you whose intellect can follow in such footsteps … but they will be few in number. For all the rest, it will be enough to have acquired the means to carry out with greater intelligence and, therefore, with greater pleasure, work which is rendered less arduous by the combination of knowledge and skill. (1825c)

After this preamble and a preliminary note came the geometry course itself, starting from scratch and extending as far as the sphere and the cylinder, with notions on the curves of lines and surfaces, the osculatory circle, developable surfaces, conic sections, and a little descriptive geometry. The work is couched in simple and clear terms, and Dupin, as usual, constantly sought to establish the relationship between geometry, practice, and the crafts. Thus, with regard to the cylinder, he explained,

> I am going to describe a practical method in the Chatham Arsenal (England) to produce, in lead, hollow cylinders of a given thickness and diameter [...] the plumber and the organ maker produce their work in the form of cylinders [...] the famous Monge, made, of the properties which I have just enumerated, a very fine application to stone-cutting.

Dupin used familiar examples. However, it is probably true that despite all his attempts at clarity and simplicity, much of what he said must have gone over the heads of his audience.

13.5. THE *MÉCHANIQUE* AND THE *DYNAMIE* COURSES

Volume 2 of Dupin's course (1825b), entitled *Méchanique*, was produced a year later, in 1826. Like the first part, it started from scratch with illustrations of the basic concepts, strengths, simple machines, ropes, suspension bridges, harnesses, ship-rigging, friction, and flexibility. Again, Dupin drew on his experiences in Britain:

> There is a type of road of which the iron rails are completely flat, with no rim, no flange, which can be simply embedded in an ordinary road or paving, flush with the road or paving. Such a system is particularly suitable for town traffic ... They have used them in the town of Glasgow for the long slope leading to the basin of the Forth and Clyde. (1825b)

Volume 3 of the course was entitled 'Dynamics' and also appeared in 1826. Dupin explained, 'I call dynamics, the science whose aim is to examine the product and the application of motive forces'. He touched on the most varied questions—the power of man, the senses, the power of animals, perspective, optical illusions (with familiar examples), the force of weight, the equilibrium of floating bodies, and the flow of fluids—and all in an extremely simple way.

13.6. Dupin's Other Activities in Education

Dupin's name appears in connection with other educational establishments apart from the conservatoire: the Ecole spéciale de commerce et de l'industrie, public courses in geometry and mechanics in the provinces, and the Ecoles d'arts et métiers. As regards the Ecole spéciale de commerce et de l'industrie (which would become the Ecole supérieure de commerce de Paris), Dupin was involved in its creation in 1820 as a member of the Conseil de perfectionnement (the Council for Improvement). Also involved were Jean-Baptiste Say, his colleague at the conservatoire; Jean-Antoine Chaptal as the president; and the councillor of state Héricart de Thury and the banker Jacques Laffitte as vice presidents. At first, the teaching was very elementary (limited to courses in writing and trading). During a session of the Conseil de perfectionnement, Dupin referred to the application of geometry and mechanics to trade, subjects that were included in the course on trade.

As regards the *écoles d'arts et métiers* at Châlons-sur-Marne and Angers—the only two that were in existence when he spoke in 1831—it would seem that Dupin, who had been a député since 1827, defended them less as a politician than as an economist, an educator, and a professor. The question was whether to end the schools or continue them, a subject that brought Dupin and Arago into conflict with each other:

> After being subjected, during the Restoration, to a strong religious influence, the schools passed, once again, to a military regime. The conservatives agitated for their suppression. Extraordinarily, Arago, who belonged to the republican opposition, fell in behind the conservatives. He demanded an end to the schools and their replacement by ten preparatory schools, situated in the main towns of France and concentrating on practical work. His aim was that they should provide excellent workers and not future captains of industry. (Delatte 1982)

This debate had already lasted for several years. For instance, on 26 June 1824, a budget commission report for 1825 was presented to the Chamber of Deputies.[4] The commission proposed to reduce the expenditure of the Ministry of the Interior by 61,000 francs, partly by making cuts in technical education (see Mertens 2004, 154). There were serious doubts about the *écoles d'arts et métiers*, because many people considered that there were enough technical schools and mechanical models in the larger towns of France. Dupin was in England at the time, during his sixth and last journey to that country. From there, he claimed that there was no regular technical education in France apart from that provided by the few *écoles d'arts et métiers*: 'Where are the towns swarming with good models, steam engines, steam boats, hydraulic presses, metalled roads, looms set in motion by a non-living force? Please name just one example!' (Dupin 1824d, 125). In England and Scotland, technical schools and mechanics' institutes were being established in many large towns, equipping their own libraries, and this, Dupin believed, was what France needed.

From London, he defended the *écoles d'arts et métiers*:

> The écoles d'arts et métiers have already rendered great and essential service to French industry; they have given to many workshops and factories, directors, managers, overseers, combining with their manual skills theoretical knowledge which it is very important to spread, for the informed practice of our arts … Let me establish as an indisputable fact: the écoles d'arts et métiers are of the greatest usefulness to French industry. (Dupin 1824d, 118–119)

The main problem, according to their critics, was that the *écoles d'arts et métiers* had failed to adapt to economic demands. The developing heavy industry required workers who were skilled in the latest techniques. Although the schools were ahead of their time, there seems to have been hesitation on the part of their teachers (Grelon 1996, 51). Should they go in the direction of apprenticeship for the crafts or turn towards a more theoretical training? Dupin again strongly defended the schools. On 11 February 1831, he wrote to 'M. le Duc' [*sic*],

> Just the day before yesterday, I received the report on which the *Conservatoire* council must today deliberate. I find it beyond

> belief that it be proposed to this Council, the natural protector of the *Ecoles d'arts et métiers*, that the two schools at Châlons and Angers should be destroyed; M. de Corbière[5] himself would not have dared to follow such advice.
>
> The reproaches made against these schools are exaggerated. The faults of which they are accused can easily be remedied. Like the memory of the Duc de la Rochefoucault-Liancourt [*sic*], these establishments are part of our national heritage; it would be vandalism to destroy them. I confine myself to placing in your hands my protest against the closure of the Châlons and Angers schools; I shall plead their cause with the Ministry and, if necessary, in the *Chambres*. (CNAM; PJ 82)

Dupin signed the letter as 'Baron, Charles Dupin, *Député de la Seine*, Professor at the Conservatoire' (CNAM; PJ 82). True to his word, he was to plead the cause of the schools at the Chambre des députés.

Two months after this letter was written—that is, in April 1831—the Chamber was dissolved, and a new Parliament was elected in July. It was in this new Chamber that Arago vehemently attacked the two *écoles d'arts et métiers*, and similarly, it was there that Dupin on 18 November responded to Arago's proposition to destroy them (Dupin 1831a). The vast majority of the Chamber adopted Dupin's point of view to keep the schools alive, and thus they were saved. Indeed, in 1827, a new organization proposed the establishment of a complete industrial programme (see Grelon 1996, 52).

13.7. The Public Courses in Geometry and Mechanics in the Provinces and Ports of France

In February 1821, two months after the opening of his conservatoire courses, Dupin envisaged making a visit to all of the naval establishments in France in order to complete (after a visit to various foreign ports) his plan for the improvement of French naval arts through the teaching of mechanics and geometry. For this, he had asked permission from the minister for the navy and colonies, but he was unable to pursue the matter further until several years later. With the 1825 publication of his *Cours de géométrie et de méchanique des arts-et-métiers et des beaux-arts* (Geometry and mechanics course for the arts and crafts and fine arts), he energetically advocated the creation of public courses in the provinces. In a preliminary note to the first volume of this work, he commented,

> We began this teaching in November 1824 at the *Conservatoire des arts-et-métiers*. More than 600 people, overseers of workshops and factories, craftsmen and simple workers of all ages and all professions have attended with zeal and concentration worthy of the highest praise.
>
> … In January 1825, M. Morin, a *Ponts-et-chaussées* engineer in Nevers, began a similar course, attended by more than 200 people. In July, 1825, in La Rochelle, M. Guignon de Grandval, professor of hydrography, also opened a course in geometry applied to the arts. In Metz, MM. Poncelet; Bergery, Bardin, Woisard and Lemoine, former students of the *Ecole polytechnique* are proposing to begin lessons of the same type, starting next November.
>
> In Lyon, M. Tabaraud, a former officer of the *Génie militaire*, will, at the same time, teach geometry of the arts and crafts. His Excellency, the Comte de Chabrol, Minister for the navy and the colonies, has just ordered all professors of hydrography, in 44 sea ports, naval or merchant, to teach the courses which we now publish. (1825b)

The second volume (*Méchanique*) has a foreword by the minister of the interior which is addressed to all of the préfets of the kingdom:

> Experiments whose beginnings augur well for success have been attempted, to make available to those who work in industry, or who intend to do so, courses where the simplest elements are taught, or rather the most frequent applications in the arts and crafts, of geometry and mechanics. Several towns have seemed inclined to follow this pattern, the course taught at the *Conservatoire des arts-et-métiers*, by Baron Dupin of the Academy of sciences. If a town in your *département* wishes to participate in this teaching, I should be greatly interested; and you may assure those who have funds to subsidize these courses, that they will have the

> administration well in support; in particular, it will assign chosen and trusted teachers. (1825b, 2:1)

In an introductory note to the third volume, *Dynamie* (Dynamics), the minister for the navy and the colonies enjoined all naval commanders to extend the benefits of Dupin's teaching to the main ports of the kingdom, where there were schools of hydrography, in the interests of industry in general as well as of the workers in the maritime professions. Two lessons a week would be given in geometry and mechanics applied to the arts and crafts on the model of Dupin's courses. These would take place in the evening, after working hours. Dupin's published courses would be provided for the professors of hydrography, who would be advised to conform exactly to his methods.

The founders of the Ecole centrale des arts et manufactures (see chapter 12.2) acknowledged the value of these courses in the provinces:

> As for the public courses in geometry and mechanics recently established in a great number of towns, on the initiative of M. Charles Dupin, and most completely and remarkably the one created at Metz by M. Poncelet, they are aimed mainly to benefit the working classes. (Comberousse 1879, 7)

In his 1826 *Tableau des arts-et-métiers et des beaux-arts* (Survey of the arts and crafts and fine arts), Dupin observed that already, two former students of the *Ecole normale supérieure* were offering their services freely for the teaching of geometry and mechanics applied to the arts in Rennes (Legrand) and in Douai (Chenon).

Dupin also gave thought to the Duc de la Rochefoucauld, with whom he was on very good terms, and his Ecole de Liancourt. The duke was a politician, scientist, and philanthropist. Fervent in the pursuit of technical progress and a member of the Academy of Sciences, he founded a model farm at Liancourt in 1769 (after a visit to England had given him the idea) and an *école d'arts et métiers* for the children of soldiers, which was to become the Ecole des enfants de la Patrie. In 1792, he had been exiled to England as the guest of Arthur Young. He was instrumental in the creation of the *caisses d'épargne* (savings banks) and beginning in 1800, he imported Edward Jenner's smallpox vaccination to France.

To the duke, Dupin wrote on 15 September 1825,

> If you could send me a young man of potential for just nine or ten weeks, for example, from 15 October to the end of December, this time would be enough for him to follow my course in geometry of the arts and crafts, to learn a little of linear design and return to Liancourt to give evening classes in geometry to your workers.

Perhaps he could not resist adding, 'I would keep this young man so very busy that he would have no time for debauchery'. On a more serious note, he pointed out that such lessons would be useful to both young and old men and that almost half of the audience members at the conservatoire were aged forty or more. He ended,

> I greatly desire, I confess, for the honour of French industry, that this course take place for the workers of the Duc de la Rochefoucauld, creator of the public schools of arts and crafts: it is for you to complete this aspect of your renown. (ADN; PJ 83)

13.8. The Movement Spreads

In 1826, a course opened in geometry and mechanics at Versailles, and Dupin contributed an introductory lecture with a statistical survey of the Seine et Oise département (Dupin 1826c). More will be said of this when I discuss Dupin's work as an economist. On 4 October 1827, he wrote to Jean-Baptiste Caquot, the president of the Société d'agriculture de la Marne, requesting the creation of a course in geometry and mechanics at Châlons. On 19 November 1827, in a letter to Baulmont (whose first name has not been found), the deputy inspector of post and the mayor of Vesoul (in Haute Saône), he expressed his delight that a course for the working class was to open in that town (ADN).

In 1828, when Dupin had been député for a year, he returned to the question of the geometry and mathematics courses that had been taught for three years in the various ports. On 24 August, he wrote to the minister of the navy and the colonies to ask if, during the journey he intended to make along the coasts of France to the Ecole royales d'hydrographie, where his *Géométrie et méchanique appliquées aux arts* was taught, he might present a gift to the professors of these schools. This gift would be a textbook, and it would provide encouragement for the success of the new teaching, which was being carried out by the professors of hydrography who put in so much effort without any special remuneration (ADN).

In that same year, 1828, a course in geometry and mechanics opened at Castres, and on this occasion, Dupin, the député for the Tarn, again contributed a speech (on 29 October). These courses, which were either already running or were to come, were Dupin's great passion, and he left no stone unturned to propagate them:

> In the absence of former students of the *Ecole polytechnique* and the *Ecole normale*, if it is possible to find an intelligent student, trained at the Châlons or Angers schools, with a little work, he could easily repeat the standard lessons given at the *Conservatoire* ... in Paris, and published in folders for the facility of the students. (1826a)

In a letter to the duc de la Rochefoucauld, a philanthropist who was active in propagating the education of the underprivileged, he explained that he had 'reduced his bookseller to providing for eight sous, each folder of geometry and mechanics, with illustrations, so that the poorest worker might buy in parts the course necessary to his instruction' (ADN). Dupin gave thought to the most minute details of the organization of his public courses: 'The professor found, the location obtained, there is just the heating and lighting, for the evening classes when the workshops have closed' (1826a).

13.9. More on the Conservatoire

Dupin continued to give his annual opening lectures of encouragement to his audience at the conservatoire, firing them with his own enthusiasm. His comments became increasingly political, particularly after 1827, when he had become a député, but his main concern was to raise the level of education among the working class. In his speech of 30 November 1826, (1826b) the *Effets de l'enseignement populaire, de la lecture, de l'écriture et de l'arithmétique, de la géométrie et de la méchanique appliquée aux arts, sur les prospérités de la France* (The effects of people's education, of reading, writing, arithmetic, geometry and applied mechanics on the prosperity of France), he had much to say on this topic (as was also seen in chapter 12.5):

> One began by thinking that mathematical truths were unintelligible to simple workers, because they are presented, in teaching manuals, in abstract and difficult form; it was thought impossible to make them easy and tangible; that was a mistake; it was the method that was defective.
>
> I would say to the tinsmith, the plumber, the boilermaker, the turner: when you cut on the cross a pipe, a roller, a funnel, you make an oval cut; and you, gardeners, do the same with a tracing line and stakes; suppose your oval had a length of two hundred million leagues, replace a stake with an eternal luminous ball, a sun, 1,348,460 times greater than the Earth; finally, the Earth itself, roll it on that oval, at a speed of 23,000 leagues an hour. Then, you will have an idea of the immense strength that the All-Powerful uses to move one of the smallest globes of one of the smallest worlds.
>
> That is how to make easily comprehensible to the workers the grandeur of our solar system and the order, so simple, so beautiful … of the eternal movements which determine natural phenomena.
>
> It was claimed that simple workers employed in the mechanical crafts, should be given only the least possible notions of what could develop their minds, exercise their intelligence, and facilitate their memory. It seemed superfluous, even dangerous, to teach them to read, to write and above all, to count … Even if certain aspects of teaching for craftsmen might not be immediately useful, since they had been, till now, employed like mindless animals, once given, it will soon become useful. (1826b,)

Such enthusiasm reaped its rewards, and students flocked to the conservatoire. 'Yesterday, as a week ago, the *Conservatoire* amphitheatre was hardly big enough for half the students', Dupin wrote to a lawyer friend on 24 December 1827. He added, 'I was received with even more warmth and pleasure than at my first inaugural lecture' (ADN).

The *polytechnicien* Henri-Edouard Tresca (1814–1885) commented, half a century after the start of Dupin's courses, 'Dupin was still young, well in control as a speaker, sometimes slow and persuasive, but sometimes also of the liveliest, adding image after image so that the most ignorant could grasp the exactitude of a principle' (Tresca 1873, 250).

It was also to be said in 1873, when Dupin died,

> In 1824, he wanted as audience simple workers. The success of his courses, then so new, was immense. The books of his lessons have been translated into eight languages. This noble example was contagious: similar courses were rapidly established in France and in other nations. (Charles Lévêque [1818–1900], member of the Académie des sciences morales et politiques; ADN, *fonds* Dupin 4J, bundle 2)

Perhaps Dupin's desire for popular education had its roots in his work with the navy in Corfu (see chapter 5.2). This is evidenced in his *Expériences sur la flexibilité, la force et l'élasticité des bois* (Dupin 1813a), in which he described two workers, one a woodworker, the other a machinist:

> These young craftsmen, brimming with natural talent, sought to understand my work and to follow it. It was a pleasure to see them giving it thought and rivalling me, in a sense. I had them do, on the objects themselves, geometry and mechanics, whilst they taught me a thousand practical results, through which I have realized how much, in the crafts, men with no instruction, contribute in finesse and accuracy of observation. I think that young engineers could not stay too close to their

> good workers, for they would themselves improve and become more skilled, through profiting from the type of experience, the one which they can never themselves achieve, that is born of manual work. (1813a)

Dupin would pursue this work in education of the workers until he was of a great age. Although he would be more and more involved in politics and economics, he remained an educator.

Endnotes

1. Turgot (1727–1781), known usually simply thus, was a government minister, politician, and economist who advocated freedom in commerce and industry.
2. E. L. Malus (1775–1812), a physicist, mathematician, and engineer. His name is inscribed on the Tour Eiffel.
3. P.-L. Courier (1772–1825), a French Hellenist. His writings are a useful source of information on the condition of pre- and postrevolutionary France.
4. *Le Moniteur Universel*, 28 June 1824.
5. J.-B. Corbière (1767–1853). He was one of the most virulent of the ultra-royalists, and he placed education under ecclesiastical control. In a decree of 27 February 1821, he ordered, 'The bases of education in the colleges are religion, the monarchy, legitimacy and the charter' (AP). In 1821, he became minister of the interior, but he fell from power during the Villèle administration in 1827.

Chapter 14

Dupin, Economist and Statistician

14.1. 1827, A Crucial Year

The year 1827 marked an important stage in Dupin's life; he officially entered the political arena by his November election as député for the second Tarn arrondissement (Castres), to which subject I shall return. This was also the year of the publication of his *Situation progressive des forces de la France depuis 1814* (Progress of the strength of France since 1814), which appeared in April and formed an introduction to his *Forces productives et commerciales de la France* (The productive and commercial power of France). It was followed by a truncated version that was aimed at the popular masses and entitled *Le Petit producteur français* (The French small producer). These works had been carefully prepared several months earlier for the electoral campaign.

However, in this chapter, I shall look at Dupin's work as an economist and statistician, as well as his numerous writings on subjects on which he was not always competent. He was verbose and wrote with a florid style, qualities that were perhaps surprising in a mathematician, but that was the fashion of the time, and he had become a politician. It has been seen that Dupin's writings on England from 1820 to 1824 had assumed an increasingly economic nature where that country was concerned, but in 1827, he concentrated on France and very much on economics.

14.2. THE *SITUATION PROGRESSIVE DES FORCES DE LA FRANCE DEPUIS 1814*

The *Situation progressive des forces de la France depuis 1814* (Progress in the strength of France since 1814; 1827a) was the introduction to Dupin's *Forces productives et commerciales de la France* (Commercial and productive strength of France; 1827c). He wrote,

> I herewith present the introduction to a work entitled *Forces productives et commerciales de la France*. I thus call the combined forces of man, animals and nature, applied in France to agriculture, workshops and trade. These forces are not static; they increase with the prosperity of the people and decrease with their decadence. I have tried to measure, for our country, not only their natural greatness, but the speed of their increase, a speed which must determine our expectations. (1827a,)

He claimed that he was neither an innovator nor a creator of systems; he offered no theories of his own. He was, he said, just a narrator and 'most often, a simple arithmetician'. He compared the productive forces and output of each French département, stating that 'comparative statistics is a science to be created, to meet the needs of our time'. As an aspiring politician, he added, 'Do not seek to find in this work to which party or colours my opinions belong. My party is France, and my banner's colours all those of the light that civilization sheds to illuminate its progress' (1827a).

He waxed lyrical on the returning soldiers who were once again taking up the tools of their trades, showing a different kind of courage than that required on the battlefield. He then produced a survey of agriculture and industry, spinning, paper, clock making, and pottery, followed by the navy and the birthrate, concluding that

> thus, by a contrast well worthy of note, the numerical increase in the population is less than that of all the material forces, of all products of work, and the increase in publications, representing the progressive activity of minds, is the greatest of all. (1827a)

However swift and widespread the development of physical activity and the increase in material wealth in France might be, Dupin considered that intellectual activity and the growth of literary wealth was even more rapid and widespread. This, he felt, was a cause for pleasure. He then reviewed the publications from 1814 to 1826 in theology, legislation, the sciences, philosophy, social economics, military affairs, fine arts, literature, history, travel, and other various subjects, summarized in the following table (1827a), which I have translated:

Productions by the press on	**1814**	**1820**	**1826**
THEOLOGY	4,974,788	7,867,609	23,268,420
LEGISLATION	1,371,568	6,326,652	18,605,495
SCIENCES	2,546,270	5,327,174	12,160,381
PHILOSOPHY	753,185	1,185,429	3032,191
SOCIAL	1,634,485	1,774,246	2,097,390
ECONOMICS			
MILITARY	441,510	1,026,027	1,445,982
WRITINGS			
FINE ARTS	773,099	1,202,599	1,999,560
LITERATURE	13,352,920	20,436,803	27,704,971
HISTORY	16,226,566	33,149,157	46,545,727
TRAVEL			
MISCELLANEOUS	3,600,478	2,121,251	7,699,977

Dupin did not provide the source of these statistics, and as will be seen in chapter 14.4, he would be criticized for using unreliable documents. Commenting on his findings and comparing the number of published pages in 1812, 1820, and 1826, he wrote of the 'fortunate progress of France and the change in intellectual tastes in the transition from Empire to constitutional monarchy'. France, now reduced to its former territorial limits, was publishing twice as much as it had in the Empire period, when it had achieved its widest boundaries. Dupin also found that 'works of literature, for the pleasures of the imagination', which had been at their height of popularity during the Empire, were now in second place. The new emphasis in publishing was on geography, travel, ancient history, and above all, modern history. Dupin referred to 'the generous spirit of the modern generation' but added,

> Who would believe it! Amidst this immense progress some blinkered minds ..., their prejudices dominated by their passions, have not yet abandoned hope of seeing a great nation turning back, its intellect diminishing, and its energy dying. Through the fortunate effect of our new institutions, tastes in France have lost their frivolity. Serious study has won first place. (1827a)

He wrote of 'a new maturity' in the people and referred to the

> so-called Christians, who have nothing in common [with the new movement and progress], individuals, infatuated with their native ignorance, since they have never known anything, believe they can teach us ... to undo learning. Blind as you are, I pity you! Let us look at the truth. (1827a)

Dupin wrote of the work of industry,

> Let us point out the services rendered by industry to the instruction of the working class. Greatly daring, manufacturers have shown ... the alphabet to their apprentices within their factories; they have resisted considerable hostility from outside, and have not closed their benevolent schools, mutual as they are. (1827a)

Of the hostility Dupin mentioned, Ponteil observed, 'The Church, hostile to the mutual method and the hand of the state on schooling, resists them both. She reproaches the former with republican and protestant tendencies. Bishops and priests warn of the dangers' (1966, 192).

Dupin also referred to military instruction as an element of civilization: 'Soldiers return to their villages with learning they would never have otherwise acquired'(1827a), he stated, and officers recognized the value of instruction for the efficiency of the military forces.

In the foreword to the *Forces productives et commerciales de la France*, he wrote,

> I began drafting this work in September 1826 and did not finish it until 1827. During that time, several branches of the national fortune have experienced enormous difficulties. French industry, so highly active, so prosperous and worthy of being so has fallen into a sorry state. Imprudent speculation, here and amongst others abroad, in England, for example, has produced a worrying counterblow for French industry and production. This dreadful state of affairs can only be temporary, but political alarm must not serve to render it more deplorable. (1827c,)

The political situation in France was worrying indeed. Between 1827 and 1830, there was a series of economic downturns and a rise in liberal opposition in the Chamber of Deputies that was ultimately to fell the conservative Bourbons. Industry and agriculture were in decline, as were the grain harvests. There was the scandal of the émigrés demanding reparation and compensation; on 21 April 1825, the Chamber of Deputies voted for one thousand million francs to be shared over five years by fifty thousand nobles, to the disgust of the majority of the population. There was a plan for the reestablishment of the *droit d'aînesse*, the feudal principle of inheritance by the eldest son, which prompted Dupin *aîné* to action and a reply (A.-M.-J.-J. Dupin 1826). In the face of growing hostility to the regime, Villèle took extreme measures, such as creating a draft bill to destroy the political press and impose very restrictive measures on the press in general. However, with the mounting opposition, he had to abandon his plan for the partial restoration of the *droit d'aînesse* as well as his proposed laws on the press. He attempted to strengthen his position by holding new elections at the end of 1827, but he was defeated and retired from the scene in 1828. More will be said of this when I look at Dupin's role as a politician. The continuing opposition, dissolution of Parliament, and ensuing bloodshed have been described in chapter 12.6.

This was the climate in which Dupin was writing his *Forces productives* (1827c). Continuing the foreword to this work, he dwelt long on public courses and the speed with which they had progressed, even while he was writing this book: 'When I was writing page 61 of the first volume, there were but 100 towns in France with courses in geometry and mechanics applied to the crafts. There are now 110.'

He then addressed the captains of industry:

> And you, successful industrialists, who come to the capital to display the treasures of your ingenious fabrication, bring us even more valuable treasures; bring us the evidence of gifts you have distributed amongst the working class. Tell us which schools you have founded for them ... and we shall not simply say of you, 'They are to be honoured for their skills', but we shall add, 'They are to be praised for their virtues and their humanity'. (1827c)

Dupin dedicated this work to the inhabitants of southern France, the Midi, explaining the extent to which they were underprivileged in comparison with the north. This must have angered the government, but it would serve Dupin well in his next election campaign, as material addressed to a southern département.

As mentioned, the introduction to this work was *La situation progressive des forces de la France depuis 1814* (Dupin 1827a). The work itself, the *Forces productives et commerciales de la France*, was in seven volumes.

1. Evaluation, dénombrement des forces (insuffisance des moyens employés à ce jour). Dénombrement de l'espèce humaine; évaluation de la force humaine et de la force animale applicables aux travaux utiles et recensement des forces infudtrielles.

[An evaluation and enumeration of forces (inadequacy of the means employed today). Enumeration of the human species; evaluation of human strength, and animal strength applied to useful work and a record of industrial strengths.]

2. Améliorations applicables aux forces physiques et intellectuelles de l'homme et de la femme. [Improvements applicable to the physical and intellectual strength of men and women.]
3. 4 and 5. Description des 32 départements du nord de la France et forces productives et commerciales de ces derniers. [A description of the 32 *départements* of northern France and their productive and commercial strengths.]
6. [Perhaps the most interesting and contentious item] Parallèle entre la France du nord et la France du sud. [A parallel between northern and southern France.]
7. Circulation intérieure de la France du nord et spécialement sur le canal maritme de la Seine, sur Paris port-de-mer et sur le canal de Paris au Rhin. [Internal traffic in northern France and especially on the Seine maritime canal.]

These sections added up to an imposing compilation.

In writing of the Nièvre department, Dupin concentrated on his hometown of Varzy, describing it as

> a town of 2,658 inhabitants, and my native place, where they make a great deal of linen. Varzy also had a pottery, abandoned today, but surely to be re-established. This town, as well as Clamecy, will certainly experience a new upsurge, thanks to the route, at last completed, linking Auxerre and Nevers.

He referred to the work in the iron industry of one of the first *polytechniciens*, Georges Dufaud (1777–1852), a man who was from his own native region. Like Dupin, Dufaud had been on exploratory trips to Britain, particularly to South Wales. He learned a great deal from his visits, knowledge which benefitted the new industry in Fourchambault on the Loire, near Nevers (Le Mallier 1971).

The *Forces productives* ends with a few general remarks, particularly on the capital cities of the main European states. Dupin observed that the richer the state and the more advanced it was in the arts and civilization, the more densely populated was the capital in relation to the total population. Moreover, capitals which were also seaports were even more highly populated than the others. Dupin announced what he saw as an important principle of social economics:

> Any industry able to favour the development of the productive and commercial forces of France, must, without fail, favour the development and increase of the productive and commercial forces of all the rest of the kingdom considered in its totality.

14.3. *LE PETIT PRODUCTEUR FRANÇAIS*

In the prospectus of the *Le Petit producteur français* (Dupin 1827d), Dupin explained that the work he had just published as the *Forces productives et commerciales de la France* comprised two large maps and two volumes. Its price was twenty-five francs in Paris, putting it beyond the means of small landowners or industrialists. So, he thought he would summarise the work in seven smaller books, in which the ideas that would be the most useful to the less wealthy would be available. These seven books he entitled

1. Petit tableau du progrès general de nos forces productives et commerciales (A brief survey of the general progress of France's productive and commercial powers):
2. Petits propriétaires agricoles (agricultural smallholders):
3. Petits fabricants et artisans (small manufacturers and craftsmen)
4. Petits commerçants (small traders):
5. Simples ouvriers (labourers):
6. Ouvrières françaises (French female workers):
7. Moyens les plus propres à faire prospérer une petite cité française (The best means to further the prosperity of a small French town).

Each of these small books, sold individually, would cost seventy-five centimes and could be bought from the publisher, Bachelier, on the Quai des Augustins. Dupin noted that several people had expressed the wish to buy large numbers of these small volumes to spread through the countryside and workshops. To encourage this, he offered the reduced price of fifty centimes each for orders of one hundred books.

The new work was dedicated to Count Daru, a member of the Institut and a peer of France. Dupin covered many subjects, even personal cleanliness, sometimes sounding paternalistic; however, what would be inadmissible today was acceptable in 1827. He wrote,

> I have drawn a plan, in black and white, of the state of popular instruction in the kingdom. This plan, with very few exceptions, is also a map of the insalubrity and cleanliness of the kingdom. In the white *départements*, the inhabitants of the small towns and the countryside are clean, dirty in the black *départements*. Why should we be surprised by this relationship between education and the cleanliness of the people?

Dupin's concern with hygiene seems to have been justified, as may be seen from a comment that was made about French troops during the First World War:

> The French poilus were notorious for their cavalier attitude to hygiene in the trenches. British troops hated taking over trench lines formerly occupied by the French, and one of the first duties they usually undertook was to clean up the lines. (Payne 2008)

Perrin (1983, 565) claimed that the state of ignorance and physical dirtiness was present even in 1939.

In the seven booklets, Dupin covered all sorts of subjects. He always illustrated his topics with simple and familiar examples, such as the installation of a chimney, the building of a house, and the bad taste he saw around him in French constructions. He even discussed the moral and practical education of children.

14.4. Reaction of the Institut to Dupin's Works

In 1827, the Institut reported on Dupin's latest publications. On 28 August, the reporter and zoologist André Marie Constant Duméril (1812–1870) commented,

> The latest work of Charles Dupin, entitled 'Forces productives et commerciales de la France' assumes, at this time, a considerable importance. The 32 departments of the north, of which this savant has provided the statistics, have sent their products to the Louvre, where they can all be judged and show the place they must occupy in the nation's industry (Duméril 1827,)

This was a reference to the 1827 French National Exhibition, which was held in the Louvre; it lasted sixty-two days, and there were 1,695 exhibitors.

The report reproduced some interesting statistics. Dupin had said that of 28,812 births, only 18,591 were the product of legally contracted marriages, 2,378 were illegitimate children nurtured by unmarried couples, and 7,843 were handed over to orphanages and received as lost children. He continued that it was rare in Paris for girls to remain unmarried, and therefore it was apparent that most of the 'dissolute behaviour that we deplore belongs to the working class' (Procès-verbaux de l'Académie des sciences, 28 August 1827). Dupin felt that this situation should be urgently remedied. He proposed different methods, particularly providing more widespread public education, offering easier and more affordable apprenticeships, teaching medicine to women, providing musical instruments (he does not specify to whom), and above all, encouraging men and women to marry. In 1840, 'concubinage' was still considered to be 'the habitual state of the working classes; the number of illegal unions is still horrifying' (Buret 1840, 417).

It is interesting to reflect that France now has the lowest per capita number of marriages in Europe, and its number of civil partnerships has quadrupled over the last two decades. The number of marriages fell from 416,500 in 1972 to 26,100 in 2006. Between the years 2000 and

2006, the number of civil partnerships in France rose from six thousand to more than sixty thousand. People in civil partnerships, which are officially recognized in France, now enjoy many of the same rights as married couples (Hampshire 2006).

André Marie Constant Duméril, the author of the Institut's report, continued that the main concern in marriage should be to ensure the well-being of the children. He observed that whether the couple were married or not, poverty was still a fact—a wife had to be fed. Even if one did not marry, the alternative was not necessarily a civil partnership. He added, 'We shall gladly share [Dupin's] opinion when he shows that marriage is a means of existence'. This was in 1827. Dupin did not marry until 1830.

Duméril observed that Dupin wanted priests to be rewarded at the expense of the commune if they merited it according to the view of the préfet or bishop. Priests should not receive an income, Dupin had argued, because that would taint their duties with mercenary interests. Duméril said that the Institut was not of Dupin's opinion—it believed that every service deserved a salary. The magistracy, the army, and the minister received an income, he noted, and the president of the royal court served the people as much as a village priest did. He continued that the most interesting part of Dupin's work was the parallel that was drawn between the north and south of France, a concept which was completely new and which would give a precise idea of France's situation and resources. He ended this far-from-uncritical report,

> Perhaps, at first, the author [Dupin] placed too much confidence in certain documents sent to him; it is very easy to see, in the descriptive section, the areas he did not visit. So, we advise him to travel further, to see for himself, to study in depth the concerns of the West and South to acquaint himself on site with the barbarous people untouched by civilization. With his gift for observation and his profound knowledge, M. Dupin will then complete, in a manner worthy of him, a work already on the front row of all our libraries. (Procès-verbaux de l'Académie des sciences, 28 August 1827)

Dupin's work with mapping was a new departure for him, and it has been considered worthy of attention in the twentieth century:

> In the beginning of the 19th century, thematic cartography was enriched with subjects related to the human world. In particular, the enthusiasm for statistics brought into the foreground several topics drawn from what we would call today 'social sciences', demography, political economics, moral studies. In this context, most of the major methods of quantitative mapping were invented within a short period, between 1826 and 1850. (Palsky 2007)

Palsky's comments on Dupin's map, the *Carte de l'instruction populaire de la France*, which was shaded according to the level of primary education in a given area, have been discussed in chapter 12.7.

Determined to pursue his work in this field, Dupin wrote on 16 August 1827 to the secretary of the Bourges Société d'agriculture with a copy of his work, a description of the productive and commercial strength of northern France. He added, 'I am now doing the same for the centre and south of France, with a description comparable to that of the north' (ADN). He asked for advice on reliable sources of agricultural statistics for the Bourges area and what types of agriculture had most particularly improved there since the beginning of the century, especially since 1814. He inquired further as to the nature of the agricultural trade in vegetables and animals.

In volume 20 of the *Mémoires de l'Institut* (1827) are three reports relating to the *Petit producteur français*. Two of these contain the prospectus written by Dupin, but the third is an extract from the *Constitutionnel* of 23 September 1827 (no. 266), describing the work as a small book which

> may become an important work and rival in usefulness much vaster compositions. This is the nature of a small volume by Charles Dupin. This benevolent economist seems to have devoted his time and his intellectual efforts to the instruction of the working classes. He publishes, in close succession, a series of treatises conceived and written on the inimitable model left by the illustrious Franklin in his *Science du bonhomme Richard*.[1]

> Much talent is needed for simplicity without triviality, measured terms without weakness, clarity without detracting from depth. These combined merits which are becoming more than ever rare in an age of pretention and swollen headedness, are indispensable, especially in a work directed at simpler minds. M. Dupin possesses them to an admirable degree.

This was high praise indeed. However, one should bear in mind that the *Constitutionnel* was edited by Dupin's brother, Dupin *aîné*.

Dupin still pursued publicity, and he presented copies of his *Petit producteur français* to various organizations. For example, he offered one to the Société de la morale chrétienne (the Christian Morality Society), claiming that the volume's aim was to improve the physical and moral state of the small landowner, small producer, small trader, and simple worker (ADN, letter of 17 September 1827). A few months later, he sent volumes 4 and 5 to the same society, mentioning that he was presenting the idea of popular libraries and useful schools for small traders and young people who were destined for trade, even though their parents were of only modest means. In the fifth volume, he pointed out, he had published various lectures on the teaching of sciences applied to industry (ADN, 2 January 1828).

Following Dupin's dispatch of the work to the president of the agricultural society of the Eure-et-Loire, he received a letter dated 12 December 1827 in which the president expressed his society's conviction that the distribution of the work would be most useful to the agricultural workers. The president reported that he wanted to see a copy in the hands of many of the workers, but the funds were lacking. However, he was pleased that Dupin was concerned with agriculture and promised to provide him all possible help in this 'generous enterprise' (ADN).

14.5. Dupin, Statistician?

I shall now look at Dupin as a statistician—a word and a science that were increasingly gaining importance at that time. As has been seen, the year 1827 marked a turning point in Dupin's life, with his entry into the political arena and his election as the député for Castres (Tarn).

It was in this year that he published his *Situation progressive des forces de la France depuis 1814*, containing choropleth maps (see chapter 12.7), thematic maps in which areas are shaded or patterned in proportion to the measurement of the statistical variable that is being displayed, such as population density or per capita income. It is an easy way to visualize variations in a geographical area.

Palsky commented:

> The process [of the choropleth map] was inaugurated at the beginning of the 19th century, by the Frenchman Charles Dupin (1784–1873). This Polytechnician, author of works on political economics and social statistics, presented the first choropleth map in 1826, during one of his lectures at the *Conservatoire des arts et métiers*. This map was published in the following year in the work *Forces productives et commerciales de la France*. Dupin illustrates in his map the theme of primary education, evidence, according to him, of the contrast in development between northern and southern France, on a line from Sant-Malo to Geneva … The choropleth map soon became widely used by statisticians and then, after 1850, by geographers, particularly in France and Germany. (1990, 5–11)

Stendhal wrote of the map,

> M. Dupin divided the map of France by a line extending from Geneva to Saint Malo, and he drew a comparison between the different degrees of civilization and information existing in the two portions of France situated on the north and south of this imaginary line. He exhibited a map of France, the curious appearance of which powerfully excited the attention of his numerous auditors. It was shaded light and dark, according to the degree of ignorance prevailing in each of the eighty-six departments. The departments in which there is least information, were of so dark a colour as to be almost black; as, for example, the department of which Tours is the capital, and in which so many English families have taken up their residence, to acquire, as they say, the pure pronunciation of the French language. The departments which send most pupils to the public schools &c. were left nearly as

> white as the paper. Unfortunately, these light parts are very few in number, and the map altogether presents a very sombre aspect. (qtd. in McWatters 1993, 7: 32)

Chartier (1992, 738) commented on Dupin's brilliance in defining the Saint-Malo-Geneva line, not so much in illustrating the differences in education between the two halves of France, but in providing a clear and straightforward use of statistics and economic indicators. Karila-Cohen (2009, 130) commented that Dupin tried, in 1828, to use the same method to explain the formation of political opinion in his *Forces électorales de la France en 1827* (the year of his election). He described the relationship between productive power and electoral power and the division of opinion, drawing on administrative and public papers to illustrate that relationship. Dupin constantly used the term 'political statistics' at a time when, between 1820 and 1830, according to Perrot, it 'was raining statistics on all subjects' (1976, 125).

As Falguerolles (2009, 117) pointed out, the appearance of the actual term 'statistics', first in Germany, then in England, France, and Italy, dates from the second half of the nineteenth century. However, it is quite possible that Dupin had read the works of the precursors of the science, Sébastien Le Prestre de Vauban or Jacques Necker. Also, engineers were accustomed to quantifying necessary materials in tabular form. Dupin would use statistics later on; in 1836, he wrote an account of the price of grain for the French population that was very much a statistical analysis. During a reorganization of the conservatoire (see chapter 13.4), Dupin's chair in geometry applied to the arts became a chair in geometry applied to the arts and statistics.

The statistical side of Dupin's work is self-evident and was recognized during a large part of the nineteenth century. However, like so many of his achievements, it disappeared from view after 1870, and after 1900, it was rarely mentioned until towards the end of the century, when writers rediscovered him. His interest in economics was apparent, not only in technical education but also in world exhibitions. He was the author of the general report that was requested by the committee for the 1834

Paris industrial exhibition (the first in a series of similar events during the July monarchy, as Louis Philippe's reign is known), and he presided over the French committee for the Great Universal Exhibition in London in 1851.

From London, on that occasion, and to return to Dupin's family life, he wrote to his daughter Marie on 17 May 1851,

> I am taking advantage of the opportunity offered me … to thank you for your charming letter; it is written in very good English, with no spelling or grammar mistakes, and it is very well thought out … I was invited to a public dinner, where we stayed at table for five and a half hours. We digested, more or less, twenty-two speeches, including my own, in English, which was received with much pleasure. I hope that Charles[2] is working. Advise him well, how he hates that! (ADN; PJ 80)

Dupin added a postscript with good wishes for the Augoyat family (see chapter 3.1).

Endnotes

1. The reference is to Benjamin Franklin (1706–1790) and his *La science du bonhomme Richard et conseils pour faire fortune*, published in Paris in 1857. Franklin began this work in 1733 as the *Poor Man's Almanack.*
2. Dupin's son, then aged sixteen. He died at the age of thirty-three, leaving no children.

Chapter 15

The Economic Situation in France Around 1827 and the Fall of Villèle

15.1. Conditions in France

At the time that Dupin went into politics as a député and had just published his first works of a truly economic nature, France was in a poor state politically, and the general situation was to lead to the fall of Villèle, as has been seen. Conditions in the economy were no better, largely as a result of a trade crisis in Britain. In that country, there had been a sudden fall in purchasing power that affected the whole of Europe. France was the first country to suffer, being Britain's best customer; this resulted in an adverse reaction to the government of the Restoration and unleashed protests, particularly in the wine trade. The producers complained of the duty that was imposed on their exports as reprisals for the duty that was levied by the French customs on, for example, the import of iron. The historian Lavisse (1921) pointed out that Dupin summarized the attacks on the protectionist system in his *Petit producteur français* and in various speeches to the Conseil de perfectionnement of the Ecole spéciale du commerce. The fall of Villèle in 1827 saw an end to the protectionist policy, and the new Chamber was very much in favour of free trade.

Dupin played a major part in the debates on free trade that took place from the end of the Napoleonic wars until the Second Empire, but his attitude varied. According to Sherman, most historians of France's experience 'have explained that French officials were really guided by their own economic class interest when they enacted policy' (1974, 1). However, Duvoir maintained that Dupin would have claimed, 'It is not I who change, but governments!' (1901, 175). Todd saw three stages in the development of Dupin's economic thinking: 'a *liberal* moment during the Restoration, *national* at the start of the July monarchy and *colonial* from 1840 to 1860' (2009, 177). Although it is true that governments and conditions changed, Dupin became increasingly ambitious, and this probably influenced his economics.

At the time of the Restoration, Dupin was a liberal, and as has been seen, he championed Monge and Carnot when they were threatened by the new royalist regime. His liberalism extended to freedom of trade, and in 1822, he spoke of the value of international trade as an alternative to war (Dupin 1825b, 77–79, 256–276). Todd described Dupin as being 'close to several liberal economists, beginning with Jean-Baptiste Say' (2009, 178). He was convinced of the value of productive labour and development as a means of breaking with France's Ancien Régime, and in his *Le petit commerçant* (The small trader; 1827d), he poured scorn on the principle of prohibiting imports. On 13 July 1827 at the Ecole spéciale du commerce, he again referred to the errors that were committed by protectionists:

> Men who make a career out of providing trade with theories, have concluded … that trade must pursue a new aim … to buy as little and sell as much as possible … If this idea had suddenly taken root … we should have seen the various nations of the globe solemnly declare that they had decided to sell forever and never buy. (qtd. in Charléty 1921, 297)

The transition from protectionism to free trade was not easy, but it was possible to improve the conditions of exchange and production within the country. To this end, there was a degree of unity between the two camps for the construction of roads, canals, and iron roads for bringing raw materials to the factories. They also agreed on the need for a more liberal relationship between trade and finance as well as an emphasis on technical education to train better workers, aims that were very much those of Charles Dupin.

15.2. From Free Trade to Neoprotectionism: The Background to Dupin's Economics

When Dupin arrived on the scene of economic study, the camps were divided between the champions of free trade and protectionism. It is not surprising that the main proponents of free trade were British. Britain, as the workshop of the world, had little to fear from competition; according to one analyst, 'On the whole, France was still a country of peasants when the industrial revolution was taking place in Britain' (Billiet 1825, 24). French industry tended to be smaller and more family controlled than was industry in the United Kingdom and Belgium, and what industry there was was more dependent on government protection and was largely intent on surviving rather than on expanding (ibid.).

French protectionism began early. Jean-Baptiste Colbert (1619–1683) had called for the enrichment of the country by commerce. New industries were established, and protections were provided for inventors. Foreign workers were invited in, and French workers were forbidden from emigrating. Colbert had the quality and dimensions of each article fixed by law, and there was severe punishment for breaches of the regulations. The constant and strict supervision wasted time and restricted improvements, but Colbert firmly maintained the corporation system, so industry remained in the hands of the privileged bourgeoisie. There was little chance of advancement for the lower classes.

In 1664, however, Colbert recognized certain drawbacks of his policy, and he reduced the import duty on raw materials that were necessary for French manufacturing as well as the export tax. However, in 1667, a heavy duty was imposed on manufactured products from abroad. In 1688, Claude Le Peletier banned the import of textiles from Holland and also of herrings, unless they had been prepared with French salt. The Dutch responded with a prohibition of imports of French wine and alcohol. Dupin saw some merit in Colbert's encouragement of industry:

> Colbert … returned life and activity to trade; furthered the interests of the nation's industry through wise encouragement; he looked at branches of foreign industry which were then flourishing, and brought it to ours. He undertook to bring back to our soil the sources of wealth and prosperity. (Dupin 1821c, 12)

In France, François Quesnay (1694–1774), the king's physician, an economist, and the author of the *Tableau économique* (Economic survey), founded one of the first schools of economics, the Ecole des physiocrates. In 1758, in his *Tableau économique*, he preached free competition and expounded the theory of the net product—that of agriculture—convinced as he was of the existence of natural laws, which he believed were the best possible. He saw the world as being divided into three social classes—the productive class, the landowning class, and the sterile class of industrialists and business.

Vincent de Gournay (1712–1759)—a business administrator to whom Baron Anne Robert Jacques Turgot (1727–1781), the economist and statesman, felt himself deeply indebted—preached plutocracy and the paramount nature of the owners of wealth. Implicit in this concept was that landowners should have complete authority over any industry that was established in their territory. Himself from a wealthy merchant family, Gournay advocated the freedom of industry and the suppression of regulations and monopolies. He was the author of the physiocrat slogan 'laisser faire, laisser passer'. In his view, the earth was not the sole source of wealth. Above all a businessman, he was close to the

Adam Smith school and considered that the products of industry and trade were also factors in the production of wealth.

Turgot, a disciple of both François Quesnay and Vincent de Gournay, was, in economic matters, a liberal. He was the author of *Réflexions sur la formation et la distribution des richesses* (Reflections on the production and distribution of wealth; 1769–1770). In this work, he recognized the dependence of the wage-earner:

> The wage of the worker is limited, by competition amongst workers, to his subsistence. He earns only his living. [The manufacturer] pays him the least possible; as he can choose amongst a large number of workers, he prefers the one who works most cheaply. (chapter 6)

Turgot pointed out that agricultural workers are productive, but those people who sell their labour receive only wages. Agricultural landowners produce more than they need and can therefore pay for labour. Like Quesnay, he saw that society now had three classes: agricultural workers, artisans and workers, and landowners. Within the industrial class, there were three subdivisions: entrepreneurs, capitalists, and simple workers.

In 1774, Turgot secured free trade in grain. Two years later, he abolished the corporations and replaced the *corvée royale*[1] by taxation. As a result of pressure from the discontented privileged classes, he was dismissed by the king, but the authoritarian and protectionist doctrine of Colbert, the French minister of finance from 1665 to 1683 (see Ames 1996 and Clément 1846), had been destroyed.

Billiet wrote that under Louis XVI,

> the arts and crafts were subjected to rules which stifled the progress of industry. Turgot decreed the abolition of all the rulings of the various corporations. There was sudden and total freedom from what had been a kind of slavery. (1825, 37)

Billiet then praised what he described as

> one of the finest institutions that the ingenuity of the administration has ever conceived, [which] makes a powerful contribution

> to the enlightenment and inspiration of our industry; I refer to the *Musée des arts mécaniques*, a vast establishment where all useful models are preserved, all inventions likely to instruct the workers are placed before them, and they are free to pursue their study of them (1825, 37).

This was a reference to the Musée des arts et métiers, which was established in 1794 with the founding of the Conservatoire des arts et métiers.

15.3. Adam Smith and the British Economists

In broad terms, Adam Smith (1723–1790) introduced the notions of capitalism, self-interest, and individual efforts to benefit society. He believed that government should not interfere with trade and that restrictions stifled competition. Jeremy Bentham, the English jurist, philosopher, and legal and social reformer, and John Stuart Mill, the English economist and liberal thinker, were in the opposite camp, believing that laws and government should help the poor and improve their living conditions.

On the eve of the Industrial Revolution, the Scottish historian and economist David Hume (1711–1776; see Henderson 1966) rejected the idea that a nation can achieve greatness as an agricultural utopia. He held that strength lies in productivity and that trade benefits everyone, bringing with it growth in the arts, as a compatible and complementary result. A country cannot become rich by beggaring its neighbours, and Hume therefore advocated free trade in advance of his pupil and friend Adam Smith. However, Hume was more cautious than Smith, favouring moderate reforms. For Smith, the Scottish economist and advocate of liberalism, the source of wealth was work, and governments should grant complete freedom to national production and external trade. He condemned the mercantilist policy of the Colbert type and demanded nonintervention by the state in economic matters and free exchange.

It was in the spirit of Smith that just prior to the French Revolution, in 1786, the Eden Treaty was signed between Great Britain and France, named after the British negotiator William Eden. It brought to an end, albeit briefly, the economic war between the two countries and established a system to reduce tariffs on goods that were imported from either country. The British prime minister, William Pitt the Younger, was considerably influenced by Smith's *An Inquiry into the Nature and Causes of the Wealth of Nations* (1776). The treaty was heavily beneficial to the British, and the unequal protection of certain industries was harmful to the French economy. British industry was so far in advance of the French that it was inevitable that French products would be disadvantaged, with consequent ill effects for the French workers. Disquiet ensued, but it was possible that French industry would soon overtake that of Britain because its labour force was so much cheaper. One writer commented:

> In 1790, business was severely affected and minds were disquieted by the effects of a commercial treaty between France and England. Since the treaty, we have proof of this. Several English textiles have been introduced to France, textiles we do not know. They have already been perfectly well imitated at a price 20% lower. (Butenval 1876, xv)

The treaty collapsed in 1793 during the Revolution. As Billiet put it, 'In year 2, all products of British industry were prohibited. At that time, France was beginning to develop the spinning and cotton industries. Industry progressed during the Revolution' (1825, 43). Billiet commented on the fine mathematical instruments that were being produced by France at a lower price than they were in London; textiles and tanneries were also beginning to thrive as a result of scientific discoveries and achievements.

By 1807, there were new guarantees and capital investments for industry, whose efforts doubled. In 1814, as a result of demands from manufacturers, there was a total embargo on imports, whatever the source. However, as Billiet (1825, 152) pointed out, such a situation results in high prices and complacency. He claimed that prohibition isolates people from each other. Rivalry and competition are healthy, he acknowledged, but total freedom of trade also has its drawbacks; he recognized that the restrictive laws had been positive for French industry.

England had recently reduced its tariffs and become very liberal in trade, not to favour foreign trade and industry but for political reasons, to foster relations with other countries, such as Latin America. British industry could not meet the country's demands, but it had a vast empire to provide raw materials. These it could process and export at a profit.

The economists Thomas Malthus and David Ricardo were broadly in agreement with Smith that the poor should help themselves and not rely on government. As mentioned, Bentham and Mill were in the opposite camp: they believed that industry should benefit the workers and improve their living conditions. Ricardo (1772–1823) became interested in economics through reading Smith. He had also researched the views of Malthus (1766–1834), who was known for his theories on population control and is familiar to many people today because of Aldous Huxley's 'Malthusian belt' in *Brave New World*. Malthus argued that as wages increase within an economy, the birthrate increases and the death rate decreases. However, during the late nineteenth and early twentieth centuries and ever since, wages have increased and the birthrate in the developed world has fallen. Malthus assumed a constant demand for labour, ignoring the effects of industrialization and automation. His was a pessimistic perspective as compared with Smith's ideas of a harmonious and stable equilibrium (Hollander 1997; Dyer 2007).

Ricardo's best-known work is his *Principles of Political Economy and Taxation* (1817). He was deeply opposed to protectionism, especially for agriculture. A member of Parliament, a businessman, and a speculator, he acquired a great personal fortune. His main contribution to the study of economics was the theory of comparative advantage, a basic argument for free trade; he maintained that there is reciprocal benefit to be derived from free exchange (see Hollander 1979). Ricardo was more interested in the distribution of wealth than in its production. He pointed out that the cost of production from fertile land was less than that from poor land, but the products were sold at the same price, thus further favouring the owners of fertile land.

John Stuart Mill (1806–1873), the author of the *Principles of Political Economy* (1848) and a dictionary of political economics (1852), sought to apply his economic theories to a kind of liberal socialism (see Rosen 2003). He is widely considered to be one of the greatest thinkers of the nineteenth century, and he was much influenced by Ricardo and Jeremy Bentham. Bentham (1748–1832) was an outspoken advocate of law reform, universal suffrage, and the decriminalization of homosexuality. He preached the doctrine of utilitarianism and the principle of the greatest happiness of the greatest number. Mill pursued the principle of utilitarianism, the separation of the public and private sectors, and female emancipation (Harrison 1995).

Inevitably, most of the renowned economists were British, because growth in the study of economics accompanies change and progress. The influence of Smith and Ricardo has perhaps been exaggerated, but public policy in the transition to an industrial society tended to be based on those ideas of the new school of political economists. I shall now return to the situation in France.

15.4. French Economic Thinkers

With the end of the Napoleonic period, French economists, particularly Dupin and Say after their visits to Britain, sought to introduce British techniques into France. However, although they made good advisers on the management of their country's industry, they contributed little in the way of economic analysis (see Lemesle 2001, 1). Adolphe Blanqui (1798–1854), a French economist and one of the first people to use the term 'industrial revolution', and above all, the politician and economist Michel Chevalier were aware of the social limits of the British system. Blanqui at first favoured it, but he soon began to have grave doubts; Chevalier (1837) during the 1830s expressed serious reservations, and Léon Faucher (1845) in the 1840s stressed the unacceptable limitations of the system, although he saw the American situation more positively than the British one. Blanqui (1798–1854) and Say were colleagues of Dupin's at the Conservatoire des arts et métiers as professors of industrial economics. As was seen in chapter 12.5, Say was the author of the first French treatise on political economy (1803) and popularized the ideas of Adam Smith. He felt that industrialization ensured growth without the risk of unemployment or fear of overproduction if trade remained free. The essential role in the economy belonged not to the capitalist or landowner, but to the entrepreneur.

In his *Traité d'économie politique* (1803), on the introduction of mechanization, Say wrote,

> The introduction of the most valuable new methods always entails some negative aspects; some interests are always linked to the use of defective methods and they are disrupted by the adoption of better methods. When a new machine ... replaces human labour already employed, a part of the workforce ... becomes temporarily unemployed. From this, serious arguments have been raised against the use of machines ... It would always be an act of folly to delay improvements forever favourable to humanity, because of the disadvantages that might occur at their origin ... The life of the consumer, and consequently, of the suffering working class, is improved by the lowering of the cost of the product ... Moreover, it would be a vain exercise to try to avoid the temporary inconvenience of the invention of a new machine by forbidding its use. If it is advantageous, it will be used somewhere ... In fact, by the use of machines, man conquers nature, and obliges natural forces ... to work for his purposes, and the advantage is obvious. (Say 1803, 1: ch. 7, 66)

Shortly after Dupin produced his *Cours de géométrie et de méchanique* (1825c), Say published his three-volume work, *Traité d'économie politique ou simple exposition de la manière dont se forment, se distribuent et se consomment les richesses* (Political economy or a simple explanation of how wealth is produced, distributed and consumed; 1803). Say began his work,

> We must distinguish between two sciences which have almost always been confused, political economics which is an

> experimental science, and statistics, which is just a descriptive science. Political economics as presently studied, is entirely based on facts. The phenomena whose causes and results it seeks to make known may be considered as general and constant, or as particular facts occurring by virtue of general laws, but where several laws act together and modify each other. (Say 1803, 1: xv)

With regard to the science of statistics, Say continued,

> It acquaints us with facts that have already occurred. It reveals the state of production and consumption in a particular place, at a particular time. It is a highly detailed description. It can gratify curiosity, but cannot usefully satisfy it without indicating the origin and consequences of the facts which it presents, and when it does not show the origins and consequences, it becomes political economics.

Although Say favoured Adam Smith, he did not hesitate to criticize his work. He described the *Wealth of Nations* as 'a confused hotchpotch of the sanest principles of political economics … his book is a vast chaos of correct ideas' (Say 1803, 1: xv).

In their writings on economics, Say and Dupin more or less ignored each other; there is no evidence of them either collaborating or explicitly disagreeing (Steiner 2009, 149). They both saw science applied to industry as the basis of social progress. However, on Say's death, Dupin said in his eulogy,

> All the principles established by this great economist are not equally irrefutable; not all are to the same degree applicable in prosperous societies or those in decline. In fact, all is not truth in the sphere of political economy, nor error in administrative practice. (MU, 17 November 1832, 1966)

Dupin was increasingly leaning towards protectionism. When a popular movement among the wine producers of the Gironde surfaced in 1828 for a reduction in trade barriers, he was very lukewarm in support (Todd 2008, 7). As Démier put it, 'According to him [Dupin], strong protection is indispensable to ensure rhythm in industrialization capable of resisting English pressure' (2009, 170). However, he was opposed to protecting archaic industries and saw mechanization as a basis of healthy internal competition. His protectionist views were complex, and he was concerned that monopolies would end in the impoverishment of the people. He similarly vacillated on the question of slavery. Having previously condemned the practice in his *Voyages dans la Grande-Bretagne* (1825g, 3:xii) in his determination to safeguard the French colonies, he took the lead in the campaign to preserve it during the last years of the July Monarchy (see Todd 2009, 187).

Say's successor to the conservatoire chair of industrial economics was Adolphe Blanqui. His liberalism was accompanied by the idea that it is not enough for wealth to be created; it must be equally distributed. The economists continued in their varying attitudes to workers' education. It has been seen that Dupin met Andrew Ure in Glasgow. About this meeting, Dupin commented,

> As early as the year 1817, I had closely studied, in a city of Western Scotland, the fine Anderson [*sic*] Institute, the first one established in Britain for teaching to simple workers the elements of the mathematical physical and chemical sciences applied to industrial skills. It had existed since the beginning of the century, and other British cities had not had the idea of doing the same. In 1818, I had carefully described and praised this establishment in the account I had prepared for the *Institut* of my early journeys to England, soon to be translated into English and published there. Our rivals overseas seemed amazed to learn, through the account of a foreigner, that there existed in the depths of Scotland, an establishment worthy to serve as a model for the whole of Europe. Soon London, then Edinburgh, then all the manufacturing towns of the three kingdoms, wanted to have scientific education for the workers. Doctor Birbeck [*sic*] [see chapter 10.4], founder of the Anderson Institute, seconded by the illustrious Brougham[2] and the generous support of a host of good citizens, multiplied the establishments and enabled them to prosper … under the name of Mechanics' Institutes. (1839b)

It should be noted that the movement for French workers' education dates back to the Revolution—hence, the Ecole polytechnique, the Natural History Museum, the Academy of Sciences, and the Conservatoire des arts et métiers were created. In 1794, it was decreed that some time in primary school should be allocated to manual work and that students in primary schools should be taken on visits to workshops to inspire in them a taste for the useful skills (*Décret sur les écoles primaires du 27 brumaire, an III, articles IX et X*). However, as opposed to this type of early instruction and the work of, for instance, la Rochefoucauld—with his *école de métiers* created in 1788 for working-class children at Liancourt—Dupin's aim was to encourage what today would be called 'continuing education'. He envisaged the development of industrial education to the benefit of the nation, but in 1839, he described the opposition he had met to his ideas: 'By striking contrast with development of the mechanics' institutes in England, and which was only possible in a régime like that of the Restoration, … the finance commission … attacked the education of the workers' (1839b).

It was in this climate that he had created workers' education at the conservatoire in 1824. Referring to all that he had seen in Britain, he called for French workers to rival the workers in that nation:

> Let us never lose sight of the fact that we shall be battling against an eminently industrious people, active and headstrong; against a people who will not easily give us the victory, and will seize back the advantage, if we relax for a single moment (1825b, 198)

Dupin saw his courses as destined for the industrial and economic success of France. They were to be a model for future teachers, who would then spread the teaching throughout France.

The economist C. L. Bergery (1787–1863), who has been neglected by historians until late in the twentieth century (see Vatin 2007), took up Dupin's call in his opening speech at the public courses in Metz:

> A colossus is rising before us and threatening us: the wealth and power of England are growing in a terrifying way, in a way to make us tremble for the future. They are growing because, in

> a country where industry rules supreme, because the thirst for positive knowledge is extreme, because knowledge has permeated even the simplest workers. (see Vatin 2007, 63)

This was what Todd (2009) described as Dupin's 'national' period—a time when he believed it was of paramount importance for France's industries to rival the great industries across the Channel.

Dupin's interest in workers' education inevitably involved him in politics. At first, he was a member of the liberal opposition that was struggling to preserve the freedoms that had been won by the Revolution, but after 1830, when the worst excesses of the Restoration were over, he became more conformist (Fox 2009, 8). He attached himself to the ideals of Louis-Philippe's bourgeois monarchy and then to those of the more dictatorial Second Empire.

However, for Dupin, workers' education was essential to a nation's well-being and the future of the nation. In 1825, he described his teaching as being for the working class (1825c). For him, education was an economic necessity, and in 1824, he had called on former Ecole polytechnique students to give something back to the nation and provide courses for the workers (Dupin 1825c, 2:151–152). It was in this spirit that an association of *polytechniciens* was formed in 1830. This should be distinguished from the original but short-lived Association polytechnique of 1816, which was launched by students, including Auguste Comte.

15.5. The Association Polytechnique of 1830

The aim of the 1830 association was to provide free and public education for working people, and it was created by the students themselves. Five hundred *polytechniciens* met at a banquet and formed the association to provide free lectures for the industrial and working classes on basic science. The courses included elementary mathematics, astronomy, mechanics, machine theory, and descriptive geometry. One of the members, Auguste. Perdonnet, emphasized the need for railway theory. Other active members were Sadi Carnot and Emile Clapeyron; François Arago, Pierre-Simon

Girard, and Jean-Victor Poncelet also lent their support (Grattan-Guinness 1990, 1112–1113). The movement was inspired by the new climate in the country. On 27, 28, and 29 July 1830, France finally put an end to the period of Bourbon Restoration. The school was therefore a daughter of a battle for freedom in which students and workers had fought side by side. The workers then became students, and the *polytechniciens* became the teachers. Classes were immediately organized.

From 1830, fourteen lectures were given freely each week: by 1835, the number was up to twenty-four, then to thirty-three in 1840 and down to twenty-seven in 1848. The public and free evening classes aimed to spread useful professional and technical knowledge and a sound general education. In 1848, several teachers left to form the Association philotechnique with similar aims. This was all a result of the movement, championed by Dupin, for the education of the workers.

15.6. References to Dupin

Throughout the nineteenth century, free trade would continue to have its advocates in opposition to the neoprotectionists such as, in particular the German Friedrich List (1789–1846), whose thinking much resembled that of Dupin. Originally, he had taken a liberal stance with regard to external trade, but he had become convinced of the necessity for moderate protection in the face of British competition. He admired Dupin as a scholar, a practical man of affairs, and 'a man of deep thought and experience' (List 1841, 55–56). List maintained that a developing nation must be strongly defended by an efficient protectionist system.

Frédéric Le Play (1806–1882), an engineer and economist, was opposed to the liberal optimists. In his 1864 *La réforme sociale* [social reform], he supported the authority of the state. He was to exercise considerable influence on the paternalistic movement among industrialists during the second half of the nineteenth century. His theories continue to be of particular interest to the right wing because, with his traditionalist and conservative ideas, he favoured an authoritarian government.

There have been few references to Dupin in economic treatises until the 1980s, but in 1982, Henderson studied him closely, looking also at the work of the chemist Jean-Antoine Chaptal. After recalling that List, during a stay in the United States, had been hotly in favour of the American system, Henderson wrote that people may be well aware of what List owed to the American protectionists, but his debt to the French protectionists is often undervalued:

> While List's debt to American protectionists is well known, his debt to French protectionists has sometimes been overlooked. His ideas were in no small measure derived from those put forward by a group of French economists who rejected the principles of laissez-faire and free trade advocated by Adam Smith and his French disciples, such as J. B. Say and Bastiat. The leaders of the French protectionists in the early nineteenth century—Jean-Antoine Chaptal[3] and Charles Dupin—considered that France should adhere to her traditional economics politics—one successfully pursued by Colbert, with such modifications as were necessary in view of changes in the economy that had taken place since the 17th century. (Henderson 1982, 262)

Frédéric Bastiat (1801–1875), a political economist and member of the French Assembly, was an exponent of classical liberalism and opposed to protectionism. DiLorenzo (1999) claimed that Bastiat acquired first-hand knowledge of trade and the absurdity of protectionism in his family business as an exporter.

Henderson provided an example of List's attitude to Charles Dupin. It would seem that List, while he was attending a dinner given in his honour at the Pennsylvania Society for the encouragement of manufactures (a protectionist society), had made a speech declaring, 'The days of laissez-faire are numbered even in France, since the famous Charles Dupin wrote his *Situation progressive des forces de la France depuis 1814* [published in 1827]' (qtd. in Henderson 1982, 265). From the introduction to Dupin's *Forces productives et commerciales de la France*, it is apparent that his views of the productive forces of France coincided with those of List (see Henderson 1982).

List, as has been seen, greatly admired Dupin. Henderson pointed out that whereas Chaptal had studied the period of 1789 through 1819, Dupin had concerned himself with the revival of France after the Napoleonic wars, covering the period of 1818 to 1827. The measures that were considered by Dupin to best favour the development of the *forces productives* (productive strength) in France were the same as those that were envisaged by List as appropriate for Germany. Although List's doctrine had been manifestly influenced by Chaptal's *De l'industrie française* (1819), it had not been influenced by the *Forces productives*, because List had already expounded his protectionist theories. But, if List was not exactly a disciple of Dupin's, he considered him to be a colleague whose views accorded with his own.

In 1837, the Académie des sciences morales et politiques offered a prize for an essay on the following subject: 'When a nation decides to establish free trade or to modify customs legislation, what facts must be taken into consideration to reconcile, in the fairest way possible, the interests of national producers and those of the mass of consumers?' (Henderson 1982, 267). Dupin had prepared notes of guidance for the candidates, with a number of questions:

> Should industries developed during a war … be abandoned after the end of hostilities? Was it not in the national interest to protect industries unable to compete with a foreign rival because the latter had taken first place thanks to an efficient machine of recent invention? etc.

However, the academy found no entry that was worthy of the prize. List, one of the candidates, was deeply disappointed, although his paper *Le système national d'économie politique* (The national system of political economy), subsequently published (List 1841), established his reputation as a leading protectionist. Henderson's article (1982) also contains much information on Chaptal, whose influence on List was considerable.

Thus, when Dupin was a professor of geometry and mechanics at the conservatoire, the conflicting doctrines of protectionism and free trade were personified in him and his colleagues at that institution, Say and Blanqui. It was as an economist that Dupin was mentioned in the 1900 'Men of the Year' (*Hommes de l'année*) in the illustrated book entitled *La vie parisienne à travers le XIXe siècle, Paris de 1800 à 1900* (Parisian life through the 19th century; Simond 1900), and no reference was made to his work as a mathematician or an engineer. Littré described him as a 'statistician' (1863–1873), and the *Dictionnaire Larousse* presented him as an 'economist, mathematician and politician' as well as being 'at the origin of the creation of French statistical services'.

As regards the policy of protectionism, Henderson said,

> Those who restrict their purchases from foreign countries inevitably find it difficult to sell their goods abroad. The protective system gave many French manufacturers a virtual monopoly of the home market. Their profits were assured and—with certain striking exceptions—they did not trouble to sell their goods abroad on a large scale. (1961, 106)

Dupin contributed to the cause of protectionism, albeit with certain reservations, while the professors of industrial economy, Say and Blanqui, were preaching free trade—an interesting state of affairs for the students. The protectionist instinct persists to this day in France; the French favour French products, which are given pride of place in the supermarkets. As Todd expressed it,

> France does not like free trade. For a majority of French people, the expression evokes firstly economic suffering, competition with countries on low wages, a reduction in social security, or delocalization. The French economy is, in practice, largely as open to exchange as those of other developed countries. But their heart is not in it … This protectionist culture goes back to the 1840s … This period saw the birth of the great contemporary ideologies, such as liberalism, nationalism and socialism. But the ideological debates of the time also bore on the question of international trade. A subject, a priori complex and technical, provoked discussion of unprecedented intensity among politicians, journalists and economists. Two new poles of ideas resulted: 'free trade' and

> 'protectionism'. The victory in public opinion of the second over the first forged the economic identity of contemporary France. (2008, 1)

At the time of writing this book, the European Community is expressing concern about French protectionist attitudes. As one writer put it, and I quote him verbatim 'Europe has to avoid the protectionism, because it is followed by the selfishness and that is the beginning of the end of the European Union itself'.[4]

ENDNOTES

1. Forced labour in the service of overlords.
2. Lord Henry Brougham (1778–1868), also a Scot, was involved in the creation of the mechanics' institutes in 1823. He fought against the slave trade and opposed restrictions on trade with continental Europe.
3. J.-A. Chaptal, Comte de Chanteloup (1756–1832), the celebrated chemist and economic thinker, created the Société d'encouragement pour l'industrie nationale.
4. Nejedlo Vit (known as Vitek), 'Europe Has to Beware of Protectionism', *Cafebabel Praha*, 24 February 2009, http://praha.cafebabel.com/en/post/Europe-has-to-be-aware-of-protectionism.

Chapter 16

Dupin's Involvement in Politics, the Transport System, and the Introduction of Machinery

16.1. Transport Development

On certain points, at least, protectionists and free traders were of one mind, particularly on the necessity to develop the transport system and to promote technical education if France were to rival Britain in industry and production. It has been seen (in chapter 10.2) that Dupin had been much impressed by the iron roads and canals he had seen in Britain. Britain in the mid-eighteenth century had begun to develop a vast network of canals. In the 1830s, such development also took place in Belgium and the Low Countries, Russia, and the United States. During the July Monarchy in France (1820–1830), there began a period of French investment in the communications system under the Becquey Plan, but as Wolkowitsch commented, 'The considerable efforts towards canal construction and river improvement in the nineteenth century were a sign that insufficient work had been done previously' (1960, 35).

This was in contrast to Britain, where rivers had been systematically improved and their navigability increased since the sixteenth century. Much work had already been done before extensions to the waterways were envisaged, and over the ensuing centuries, the work continued. In France, however, there had been neglect: 'The Loire, the most heavily used river in France, was littered with sandbars and suffered from an irregular flow. As late as 1787, it could take as much as eight months to reach Nevers from Nantes, using the Loire' (Szostak 1991, 56). The English agronomist Arthur Young travelled extensively in France between 1777 and 1790. After taking a ride on the Seine, Young said, 'It was rougher than I thought a river could be' (1792, 972).

As regards the French roads, they had similarly suffered. Arthur Young praised the French network highly, but as Dunham said of early nineteenth-century France,

> It is an unpleasant surprise to find that most contemporary and many modern writers were loud in their complaints of the bad conditions of French roads—that they were badly built, or badly maintained, that there were extensive gaps where they were almost impassable, and that the cost of travelling on them was high. (1955, 14)

The government of the July Monarchy invested in highways, with a resulting great improvement. The Saint Quentin Canal was built, the Rhine-Marne and Aisne-Marne canals were begun, and steam shipping developed, but France's mercantile marine could in no way rival the British merchant service. During Louis Philippe's reign, the most important work in French transport was railways. With an increasing consciousness of the importance of a railway network, industrialists were interested in its development. Engineers such as Gabriel Lamé and Emile Clapeyron (see Bradley 1981), who had become aware of railways during a stay in Russia, returned to France with plans for promoting them. Starting in 1825, the Saint-Simonian journal *Le Producteur* stressed the need for this development, and the most influential advocates of the railways were nearly all Saint-Simonians. Finance was to be a major concern, however, and as Clapeyron commented, 'It took Monsieur [Emile] Péreire several years to overcome the repugnance of the financial world where these new ideas were concerned' (Bradley 1981, 307). Henderson noted, 'By far the most significant development in transport in Louis Philippe's reign was the construction of the first railways' (1961, 110).

Unlikely as he was to miss being involved in a matter of such importance, what was Dupin's role in the changes that were taking place in transport? First, he was concerned with canals. In the Nièvre archives among the Dupin papers is an act of constitution from the director-counsellor of the Paris-Rhine canals, dated 22 June 1825. Dupin was present when the act was drafted, together with the Duc de Montmorency; the Duc de Damas; the Marquis de Nérac, pair de France; Comte Pierre-Barthollémy Portal, the minister of state and pair de France; Garnier de Dufougeray, the treasurer of the Chambre des députés; and Dupin *aîné*, a lawyer. Charles Dupin is listed as 'Le baron Dupin, member of the *Institut*'.

The meeting heard the reading of a submission to the minister of the interior which sought to obtain

> the concession in perpetuity and as a private enterprise, for a canal lateral to the Marne from Paris to Saint-Didier, and another … from Saint-Didier to Strasbourg, as well as various basins, warehouses and stores to establish on this navigable line. (ADN)

The meeting accepted the title and functions for canals from Paris to the Rhine as well as for basins, stores, and warehouses. At a future session, they would proceed to the choice of a *ponts et chaussées* (bridges and highways) engineer to be appointed and the commission to be established for the execution of the royal ordinance of the previous May (ADN, fonds Dupin, 2O J 29). As regards railway development, the same archive source reveals Dupin's participation in the project for a line in his home department, Nièvre. France lagged behind in railway building, and it looked as if the line would not materialize, but in 1842, a law was passed granting railway concessions with state participation. As Henderson said,

> Above all the 1840s saw France's first railway boom. By 1847, some important main lines were in operation such as those from Paris to Le Havre, Lille and Orléans and that from Strasbourg to Basel. Railway construction stimulated the iron industry and made it possible for French manufacturers to serve much wider regions than before. (1961, 109)

The following letter concerns the application of the 1843 law; it was addressed to Dupin's elder brother, député de la Nièvre, on 21 November 1843:

> A temporary committee has been formed in order to establish a railway from Auxerre to Nevers, passing through Clamecy; it has communicated its intentions to M. le baron Charles Dupin, your brother, who was so good as to accept the honorary chairmanship spontaneously offered to him. (ADN; PJ 87)

This letter was signed by Husquin de Rhéville from the committee's headquarters at the Rue des Filles Saint-Thomas in Paris.

As it transpired, the project was not easily implemented. An amendment presented to the Chambre des députés in 1846 and defended by Dupin *aîné* and another Nivernais, Benoist d'Azy, was rejected. Auxerre was still not linked with Nevers when the Montargis-Nevers line was inaugurated in 1850 with much pomp and circumstance by the Bishop of Nevers. It would not be until the Second Empire that Nevers would be linked with Clamecy and, from there, to Auxerre. This took place thanks to the intervention of Dupin *aîné* and Claude Alphonse Delangle (1797–1869), the minister of the interior in 1858 and 1859, who exerted such great influence on Napoleon III that the arguments in favour of a railway through the Nièvre Valley were finally accepted by the government.

As for Charles Dupin's concern with railways, in 1838 he presented to the Chambre des députés a report on the railway from Paris to Rouen, to Le Havre and Dieppe (this document is at the library of the Institut de

France). This report was drafted for a special commission that had been appointed to examine the railway project. Similarly, on 2 July 1840, Dupin presented a report to the Chambre des pairs on behalf of a special commission that had been appointed to examine the draft bill for railways from Paris to Orléans, from Strasbourg to Basel, from Andrézieux to Roanne, from Montpellier to Nîmes, and from Lille and Valenciennes to the Belgian frontier. His speech to the Chambre des pairs at its 1841–1842 session provoked a general discussion of a bill concerning the establishment of the great railways. For facilitating the mobility of labour and the promotion of industry—subjects that were close to Dupin's heart—these developments were essential. Paris is still easily reached from Nevers thanks to the dedication of many people, including the Dupin brothers, to the construction of railways.

16.2. Dupin's Entry on the Political Scene

It was in 1827 that opposition to the Villèle government reached its height. People from all walks of life rallied to support François P. G. Guizot,[1] one of 221 signatories to a document opposing the policies of Charles X and favouring a parliamentary monarchy. Guizot's supporters founded a society called 'Aide-toi; le ciel t'aidera' (God helps those who help themselves). Society members were to distribute brochures, keep a watch on the administration, and check the electoral rolls. The society was a resounding success, and within a few weeks, committees were formed in some sixty departments; some fifteen thousand voters whose names had been omitted from the register, a quarter of the total number of eligible voters—were reenfranchised through the committees' efforts. Villèle reacted promptly with draconian measures; at the beginning of November, he dissolved the Chamber of Deputies. The subsequent electoral campaign, although it was brief, was fought with extreme determination and resulted in triumph for the opposition. It was in this political climate that Dupin *aîné* (who had been a member of Parliament since 1815 and a député in the Sarthe since 1827) was elected for the Nièvre. Simultaneously, Charles Dupin was elected in the Tarn:

> On 17 November 1827, he began his parliamentary career. Elected deputy for the second *arrondissement* of Tarn (Castres) by 272 votes of 504 voters … against 222 for M. de Lastours, the defeated *député*, he joined ranks with the liberals and made his first speech justifying the epithet 'deplorable' as applied to the outgoing minister. (Robert, Bourloton, and Cougny 1891)

According to Christen (2009b, 20), Dupin envisaged changes in France on the pattern of Great Britain, including limiting royal powers, ending religious fanaticism, and favouring the progress of industry and the provision of a quality technical education for the working masses. His most important speeches in 1828 and 1829 were, not surprisingly, concerned with the navy, equipment, and construction; in them, he evoked the British model (see Benoît 2009, 57). He referred to 'this wise prudence which, during peacetime economises both arms and ships, so that they be numerous and formidable at the first sign of serious hostilities (AP, vol. 59, 33).

A few months after his election, Dupin published his *Forces électorales à la fin de 1827 et situation progressive des forces de la France depuis 1814*, which was already in its eighth edition in 1828. In fact, there was little new material in this work: essentially, it reproduced his *Situation progressive des forces de la France*, which had appeared in 1827.

16.3. Dupin and Lamartine

It would seem that the original publication of the *Forces productives* prompted a correspondence between Dupin and the poet Alphonse de Lamartine,[2] as Marie-Renée Morin found in her research among Lamartine's private papers. Of the *Forces productives*, she said,

> It united in one work all the elements of French civilization. It evoked the intellectual wealth; here figured the *Académie française*, where the author wished to see the presence of Lamartine. The latter was certainly aware of this homage to him because he read the book, as was apparent from his humorous letter to Virieu,[3] at the end of July 1828. (1995, 4)

In this letter, Lamartine wrote,

> Oh! Que Monsieur Dupin est un grand philosophe
> Quand il dit qu'ici bas chacun meurt à la fin,
> Et qu'aussi chaque siècle est fait de son étoffe.
> Qui s'en serait douté sans ce monsieur Dupin? (Morin 1995, 4)

Roughly translated, this reads,

> Oh what a great philosopher is Monsieur Dupin
> When he says that everyone on earth dies in the end,
> And that every century is made up of its own material,
> Who would have thought it without this Monsieur Dupin?

It is probable that Dupin and Lamartine first met on the benches of the Chamber of Deputies. Morin said of them,

> Both had in common youth attracted by travelling, political curiosity, an interest in foreign behaviour and both were utter spendthrifts ... Both were fired by ambition and the will to succeed. They knew how to make themselves known, to obtain and retain useful connections and to get their families and friends to work on their behalf. (1995, 4)

More will be said of this interesting relationship.

To return to the book which had prompted it, the *Forces productives*, this was not the first time that Dupin had reproduced work that had already appeared. His *Applications de géométrie* was effectively a collection of papers he had already produced (see chapter 6.6). However, such publications kept him in the public eye—a concern that, as has been seen, was a constant preoccupation for Dupin. Moreover, the *Situation progressive des forces de la France* had constituted a strong contribution to his election in 1827. He was again making use of them to further his parliamentary career.

In the *avant propos* (foreword) to the *Forces électorales* (electoral strength), he claimed,

> I wanted to show the generation of the new régime the secret of its power … Here is the eighth edition, which I publish prior to the re-elections. So that the constitutional royalists should be aware of their superiority on all fronts, I ask them to cast an eye over the *Situation des forces électorales à la fin de 1827*. That will give them confidence for the activities of 1828. (1828b)

The dedication was couched in grandiloquent terms: 'Voters, be strong, reject the men whose careers have been sullied. Reject the men who were oppressors when in power, and persecutors when they had influence'.

Referring to the 1827 elections, he declared, 'A great political event has been achieved. The will of the monarch pronounced the dissolution of the Chamber of deputies. Elected for a constitutional and royalist area, I have the honour to belong to the class of constitutional royalists' (1828b). The constitutional departments had produced more than two thirds of the votes that were cast.

16.4. Dupin as Député

The *Dictionnaire des parlementaires français* provides valuable details of Dupin's parliamentary career. Still concerned with Greece, in 1827 he spoke about the naval battle of Navarino during the Greek War of Independence and about the welfare of the Greeks. He had restored to the eminent geometer A.-M. Legendre (1752–1833) the pension he had received from Napoleon, which was subsequently withdrawn from him. After refusing the post of director of the Arts et manufactures, Dupin introduced into parliamentary debates the use of statistical results and presented to his colleagues an exposé of the effects and dangers of lotteries in various parts of France. He was a frequent speaker on matters relating to the navy and civil engineering. In 1829, he produced a discourse on the composition and election of the *conseils généraux* (more or less the equivalent to the British county councils). A supporter of the blockade of Alger (of which more will be said), he expressed the view that France should quickly take the offensive. In the spring of 1830, he was one of 221 men to overthrow the ministry of the notorious ultraroyalist Polignac

(1780–1847), the prime minister from 1829 through 1830. With the dissolution of the Chamber, Dupin was violently attacked by the ultra-royalists and on 23 June 1830, he failed to be reelected in Castres, where he received only 256 votes against 333 for his opponent Lastours, whom he had defeated in November 1827. However, he returned to the Chamber as député for the Paris tenth arrondissement (today, it is the seventh) on 12 July 1830, with 336 of 530 votes against 164 for Baron Leroy, the defeated député.

Dupin was a constant contributor to debates. He defended the 1832 budget as a government commissioner. After that, he was five times appointed to the commission of finance and was four times the reporter of the naval budgets. He involved himself in all of the laws for progress, management, and pensions in both land and maritime affairs. In politics, he was growing increasingly conservative, and he defended the clergy. After his reelection as a député in 1834, he was appointed minister for the navy. His ministry was short-lived, but he made time to institute a prize of six thousand francs for the most important progress in the application of steam for the navy.

It was at this time that Dupin was concerned with the creation of *caisses d'épargne*, or savings banks, and more will be said on that subject. On 3 October 1837, Dupin was elevated to the peerage, where he continued to be very active on matters such as money lending, commercial transactions in France and its colonies, and child labour. At first he distanced himself from the 1851 coup d'état, but eventually, he rallied to Louis Napoleon, who admitted him again to the Senate on 26 January 1852. There, he took part in most of the debates on the economy, politics, and religious matters. Over the course of ten years, he had played an active part in developing more than one hundred laws and had been a member of more than fifty commissions and a reporter for more than thirty.

In 1830, Dupin found the time to get married to Rosalie-Anne Joubert at Angers, just prior to the 1830 revolution and his defeat in the Castres election. According to the marriage certificate, which can be seen in the archives of the Angers *mairie* (town hall), no other member of the

Dupin family was present at the ceremony. Interestingly, despite Dupin's age—forty-six—his father had to give his consent to the marriage, as did the minister for the navy. The wedding took place on the day of the capture of Alger.

I shall now return to other events of 1830.

16.5. MACHINES

It was around 1830 that Dupin was involved in discussions with groups of workers on the subject of machines; in his admiration of British industrial progress, he defended the use and effects of such innovations. As he had claimed in his *Petit producteur* (1827d, 3:34), if weavers who worked at home were equipped with machinery, they could become rich and independent. Jarrige provided an example of how Dupin, when he was travelling in the south of France in 1828, aroused the wrath of the workers of Montauban:

> The people ... were not disposed to allow M. Charles Dupin to make a triumphal entry into the town, added to which, certain ill-intentioned people had spread the idea that this savant was travelling with the purpose of everywhere establishing machines which would deprive the workers of their livelihood; he would not have been able to cross the bridge unscathed. (AN F7 6772, Montauban, 21 October 1828)[4]

The spread of machine use had already sparked trouble and the destruction of machinery during the Restoration (Jarrige 2009, 225), but Dupin championed the new processes among the Paris labour force. In July 1830, Paris typographers broke up mechanical presses at the Imprimerie royale, and Dupin apparently received many letters every day from typographers who were opposed to mechanical presses. Far from supporting their cause, he issued his *Appel aux ouvriers français, pour les engager à ne pas briser les machines productives* (An appeal to the workers of France not to break the machinery of production) (published in Paris in 1830), and on opening his course at the conservatoire in

1830 before a large and mixed audience, he praised the advantages of machinery for public prosperity and the welfare of the working class. However, the working people were not convinced by his arguments, and he became a target for their attacks. They claimed that Dupin was ignorant of their conditions—arguments which were to continue for many years, along with the frequent refutation of Dupin's statistics. With all of his knowledge of developments in Britain, it is surprising that he did not refer to the Luddite riots of 1811 and 1812 and the British workers' reaction to the introduction of machinery. It has often been considered that this was a thoughtless battle against progress, but Thompson (1963) claimed that the workers were not opposed to the new technology itself but rather to the resulting abolition of fixed prices and the introduction of a free market.

16.6. Dupin and the Conservatoire des arts et métiers, 1830

Dupin made his customary address to the students at the opening session of the conservatoire in 1830 (1830b), when France was in political turmoil:

> Every day, on various pretexts, there were meetings and processions of workers culminating in marches on the ministries and the Palais Royal. These disturbances degenerated into violence with regard to the trial of the ministers of Charles X, Polignac and three of his colleagues arrested in their flight. On two occasions, when the Chambers abolished the death penalty for political offences (17–18 October), then on the day of the ministers' condemnation to detention, the disturbances almost turned into insurrection (Malet and Isaac 1922, 3:)

The *Journal des débats* (Bertin, 1789–1944) of 20 December 1830 reproduced Dupin's politically oriented speech closing speech to his students, *Discours sur le sort des ouvriers considéré dans ses rapports avec l'industrie, la liberté et l'ordre public* (On the destiny of workers considered in its relationship to industry, freedom and public order; 1831b) and commented,

> After recalling all the attention given to making the working class better educated and wiser, M. le Baron Charles Dupin pointed out the fruits of these efforts during the days of July; he ended his discourse in the following exhortation, at the moment when the riots were threatening the lives of Charles X's ministers, then on trial:
>
> 'Listen to me, in the gravest of circumstances! you know that my voice is that of a faithful friend, who hopes only to live a little, in the memories of his fellow citizens, by the good that he has been able to do for you ... What I am about to say would be, doubtless, superfluous, for the captains of French industry, for you all who come today to listen to me within these halls. But, alas! It is not superfluous ... for a large number of workers, still deprived of the basic notions of primary education [it was to be another three years before primary education was instituted by the Loi Guizot], and, unable to read or write, are ignorant of the truth that may be useful to them ... Continue to repeat to them that an unarmed enemy must be treated as sacred'.

Dupin went on in florid terms, appealing to the patriotism of his audience—successfully, it would seem, because the reporter described the resulting standing ovation. However, there were to be changes in the people's attitude to Dupin. Seen increasingly as a bourgeois philanthropist who was attaining high honours, he was losing his popularity. Eventually, unrest spread to the conservatoire students, who by 1848 were making their feelings felt within the auditorium. Dupin's publications (1833b, 1840a, 1840b), particularly those in which he called for calm in the face of growing unrest, provoked reactions in the working-class press. Facing increasing criticism from workers, he emphasized the economic arguments in favour of machines and their advantages to workers in speeding up and easing their tasks.

As has been seen (in chapter 15), Dupin relied on statistics and protectionism, in opposition to the policies advocated by Say. The workers rejected his statistics and questioned the knowledge he claimed to have of the industrial world. The *canuts* (silk workers of Lyon) had rebelled against the introduction of machinery and factories. Highly skilled and with a long tradition of creating fine products, they were now at the mercy of traders who could buy factory-made materials much more cheaply. The only alternative for the workers was toiling for long hours in the new factories (compare this situation with Thompson's comments about the Luddites in chapter 16.5). In 1831 and 1834, they rebelled, and in 1834, they were finally put down after much bloodshed. The resulting penalties were harsh in the extreme. In 1834 (see Jarrige 2009, 228), the Association des mutuellistes de Lyon wrote to Dupin,

> Such enormous ignorance of our industrial body, gives us a poor idea of your knowledge and your professorial science. It is with irony that the Lyon manufacturers, who are 'neither mathematicians nor academics', pitch their simple 'common sense' against the 'profound calculations' of the professor. (*L'Echo de la fabrique* [a workers' journal], 62, 9 March 1834)

The arguments continued in the working press, particularly *L'Atelier* and *L'Echo de la fabrique*, and Dupin's economics were denounced, as well as the taxation of grain imports. In *L'Echo de la fabrique*, it was commented,

> M. Charles Dupin, father of the workers, in order to favour the *députés*-landowners and to make people pay more for their bread, recently, with regard to the grain laws, without blush or hesitation, praised the English law on grain legislation (the corn bill) … The corn bill is the disgrace of British legislation and one of the first that the reformed parliament will attack (*L'Echo de la fabrique*, 26, 22 April 1832)

In the opinion of *L'Atelier*, only the workers enriched a country—Dupin offered nothing. In an article entitled 'Very Sincere Thanks to M. Charles Dupin for His Advice to the Workers of Paris' (Très sincères remerciements à M. le baron Charles Dupin, sur ses conseils addressés aux ouvriers parisiens), the paper made these feelings very clear:

> You, when you have six months of the year in making everyday laws, judging and condemning, holding forth at your leisure at the *Conservatoire*, or throwing onto paper a few more or less bad ideas, what profit does that bring to France? (*L'Atelier*, 13, 2 October 1840)

Condemnation of Dupin continued, not least for his attitude toward slavery. On 14 November 1834, he had briefly held the post of minister for the navy and the colonies in the Maret government. From that date, he became one of the protagonists of slavery, continually defending the right of colonialists against any onslaught on the slavery system. In 1845, when the Mackau law attempted to improve the condition of the slaves, giving them the right to possess their homes and furniture, Dupin headed the opposition in the Chambre des pairs on 3 April 1845 (Dupin 1844a). He continued to defend the practice of slavery until its abolition in 1848,[5] basing his arguments on what he claimed were the better conditions of French colonial slaves compared with those of British industrial workers. In his three 1844 papers on the colonies (1844a, 1844b, and 1844c), he decried the 'vices of the black race' and, in particular, their 'horror of regular work'. Moreover, he stated, 'Slaves born in the French colonies are less far removed from the feelings and customs of the family; they are closer to humanity; they are less depraved by barbaric ideas and tendencies' (Dupin 1844b). He saw interracial strife and economic decline as the inevitable results of emancipation.

The press continued to attack Dupin. The journal *Le Populaire*, widely read by working people, described him as 'narrow-minded, ignorant, presumptuous, foolhardy, slanderer, a pygmy. We could also well say that he is just a charlatan, an ass, vainglorious, who swaggers around as a baron' (*Le Populaire de 1841*, December 1844, no. 5).

In 1848, with Louis-Philippe overthrown and the Republic proclaimed, Dupin was elected to the Assemblée constituante, becoming firmly a member of the conservative majority. Where was the Dupin of yesteryear? The change in him was marked when he addressed the conservatoire students in 1848, as reported in *L'Atelier* ('M. le baron Charles

Dupin est toujours notre bon père' (Dupin is still our good father) on 4 January 1849:

> A murmur of disapproval ran through the workers. We shall not repeat all the remarks or the somewhat lively observations exchanged rapidly during the address of M. le baron. We saw only that, for several months, the audience at the *Conservatoire* had become more critical, and that, in order to convince them, it needed more than assertions and promises … Shall we say that his honeyed words were greeted by the most unseemly of jibes, 'We have been rocked too long by that song!' Some even called our noble professor a reactionary. Such were the thanks and the farewell from these ungrateful children.

In the next chapter, I shall look at other events of these years.

ENDNOTES

1. François P. G. Guizot (1787–1874) published widely on French and English history. In 1830, he became the minister of the interior, and from 1832 to 1836, he was the minister of education, responsible for considerable reforms.
2. Alphonse de Lamartine (1790–1869), a poet and politician. From 1830, politics and ideas for social reform became an essential aspect of his work; he championed democracy and the cause of peace.
3. Aymon de Virieu (1788–1841), Lamartine's great friend and his first publisher.
4. I am indebted to François Jarrige (2009, 223–234) for his research on the workers' press and his indications of where useful records may be found.
5. Slavery had first been abolished in France in 1794, during the Revolution. It was restored in 1802 by the Consulate and finally abolished in 1848.

CHAPTER 17

PARLIAMENTARY AFFAIRS TO 1848

17.1. WRANGLING ABOUT THE *ECOLES D'ARTS ET MÉTIERS*

In 1831, there were violent debates in the Chambre des députés on the subject of the closure of the two *écoles d'arts et métiers* at Châlons-sur-Marne and Angers. A duel of speeches took place between Dupin and Arago, the renowned astronomer, physicist, and politician. Arago headed the critics who were in favour of abolishing the schools; they claimed that these schools were ill equipped and that their students did not acquire the practical sense that was essential to industry. In opposition to the critics were Dupin and Argout, the minister of trade and public works.[1] Argout wanted to see not only the existing schools preserved but the establishment of new ones—and as many as possible. Dupin recalled all that he himself had already done for the schools and expounded on what still needed to be done, but he expressed his sadness about being in opposition to his old friend Arago. He spoke of La Rochefoucauld, the founder of the schools, and how upon his death in 1827, his coffin was followed by grateful former students of the schools. Waxing ever more eloquent, Dupin declared, 'No ..., you will not destroy these schools to which La Rochefoucauld devoted his affection and his benevolence; national glory impregnates these schools which produce so many skilled craftsmen, so many citizens of heroic heart' (Berland n.d). The schools did survive and were to become engineering schools in the twentieth century.

17.2. Some Local Matters

In 1834, the bishopric of Nevers, in Dupin's native territory, was under threat. It was one of the oldest bishoprics in France, but since the early 1800s, it had been administered as an annex of Autun. On 4 July 1821, there was an allocation of funds for eighteen new chairs, 'in towns where they were considered necessary and that of Nevers was re-established according to that law'. The reestablishment was contested in the Chamber of Deputies in 1833, but Dupin *aîné*, then the president of the assembly, rose to its defence (A.-M.-J.-J. Dupin 1853). In 1834, the attack was renewed in the Chambre des pairs, and this time, the bishopric was defended by Charles Dupin. Thanks to the efforts of both brothers, the bishopric was preserved (ADN).

In 1834, Charles Dupin became a conseiller général (the nearest equivalent to this is perhaps the English county councillor). On this occasion, he received a comparatively amiable letter from his father:

> Clamecy, 10 July 1834
> To Charles Dupin, member of the *Conseil général*, Nevers.
>
> You, my friend, will be at the *Conseil général* with your brother [Dupin *aîné*]. It is time for me to give you some paternal and friendly advice. Come to terms with him in private, so that a difference of opinion on anything at all, may not produce a collision between you, always unpleasant in the presence of others.
>
> Be sparing with the public purse.

> Encourage religion, morality, education suited to the different levels or professions, the good choice of public servants, the repression of rebels and brawlers. (DRB; PJ 86)

As has been seen, Dupin was elected in 1830 as deputy for the Tenth Paris arrondissement. When seeking reelection in 1834, he gave an account to the electorate of his work in the previous four years (Dupin 1834b). He claimed that he had defended social order and protected laws, public peace, and domestic security; that he had fought for the economy, agriculture, industry, and commerce; and that he 'would have blushed to give preference to wealthy interests, always clever enough to protect themselves'. As a defender of the working class, he had fought for the *écoles d'arts et métiers*, rejecting the proposal to close those at Châlons and Angers. He continued his work in favour of the *caisses d'épargne*, savings banks for the benefit of the working class, of which I shall say more.

17.3. Dupin as Minister for the Navy and the Colonies

Dupin was the minister for the navy and the colonies from 10 to 18 November 1834. A letter headed from the ministry dated 15 November 1834 and addressed by Dupin to a certain 'Mr. Vortier'[2] indicates that he had already resigned after a mere five days:

> Your congratulations, for which I wholeheartedly thank you, arrived when the duty of a good citizen and honourable man had already imposed on me the duty to submit to the King my resignation. I regret the good I hoped to do for the navy. (ADN; PJ 88)

The *Moniteur* of 6 December 1834 reported the reasons given by Dupin for his resignation:

> From the moment I was called to take my seat in the cabinet, insults of all kinds were heaped upon me by the so-called defenders of freedom, to which I was not enough of a coward to respond with flattery. I declare here and now, that fear of the inflammatory

> press will not throw me into the *Résistance* through hatred, nor into the *Mouvement* through fear. I shall speak according to my conscience and as a good man should.

The Mouvement was a wing of Orleanism, favouring a democratic regime, an extension of suffrage, and sympathy to the anticolonialists. The Résistance was moderate, bourgeois, and in favour of the '*juste milieu*' (middle way). Dupin attached himself to the *Tiers parti* (Third party), whose leading figure was André Dupin. The members of this party stood between the left and the right, voting either way; their policy was characterized by conciliation, forgetting the past, and appeasement.

Dupin explained that the cabinet had been dissolved because of the disagreement of Maréchal Etienne Gérard[3] with the rest of the cabinet, and that the attitude of the candidates for the new cabinet made it impossible for him to stand for reelection. Gérard's cabinet had been formed on 18 July 1834; Dupin immediately took up a position close to the *Tiers parti* and the Republicans in favour of granting amnesty to the instigators of the April 1834 riots in Lyon (see chapter 16.6). On this matter, he clashed with Louis-Philippe and had no alternative but to submit his resignation, which he did on 4 November 1834. It is noteworthy that during 1834, there were five ministers for the navy. Dupin's great friend Tupinier would hold the post in 1839 for forty-three days—brief periods of glory for both of them.

17.4. Charles Dupin, Pair de France, 1837 to 1848

In the Chambre des pairs, Dupin's activities were as diverse as was his career, and they were even more so than they had been when he was in the Chambre des députés. Some examples of his speeches, with their titles translated into English, will provide an indication of this:

- July 1838, a report to the commission on the railways from Paris to Lyon and from Le Havre to Dieppe
- 2 April 1840, 'In praise of M. le Baron de Prony' (Dupin 1840b) (Prony, who died in 1839, was one of Dupin's great friends, as was seen in chapter 3.1.)

- 1 March 1841, a report on the bill on the organization of the general staff of the navy

The '*Moniteur Officiel*' (as Guizot [1864] called the *Moniteur universel*, probably because it was the official organ of the government) of 28 March 1841 reported on Dupin's speech about the proposed fortification of Paris. 'After the Napoleonic wars, it had been generally considered that such fortifications were unnecessary, and there was widespread astonishment when such a move was proposed in 1841. However, the political climate inspired a fear of war in France against a new European coalition. Hence, a law was proposed and accepted for a system of defence to be constructed around Paris. Later, this decision was regretted, but by the time public opinion overt the project had changed, there was concern about losing face, and fourteen fortresses were erected, to the indignation of the population' (Guizot 1864, 8: 36).

In his speech, Dupin, in a vehement attack on the plan, explained the reasons for the opposition to it. He claimed that such fortifications would be a heresy, both for the financial burden they would place on the state and for the nation's morale. He tried to show how in the past, fortifications of Paris had never provided protection but rather had been a danger to the population of the city. He declared,

> The law ..., if you pass it as it is proposed, is the dawn, or rather, the twilight of a military revolution ... It is in the middle of this appalling maze of illusions, that an order appeared, on 10 September last, for the immediate commencement of works which would involve an expenditure of 140 million, in order doubly to surround the capital, firstly with 94 bastions of continuous walls, followed by a circle of external structures and fortresses. (*Moniteur universel*, 28 March 1841)

He then commented on the progress France had made in trade and exports; the Seine département had exported 100 million francs' worth of produce in 1840—'goods produced with marvellous skill in countless workshops' (ibid.).—compared with 50 million francs' worth in 1830.

Dupin was, of course, unsuccessful in his opposition to the plan for fortification, which was implemented.

In April 1842, replying to the minister of the interior on the sugar law, in a discussion about secret funds, Dupin (1842d) defended the colonial sugar producers. Between 1836 and 1843, he published six brochures demanding either the suppression of the sugar beet industry or a parity of taxation on sugar that was produced in the colonies and in metropolitan France, in the interests of French colonialists. Dupin had become a representative for Martinique in the Council for Colonial Delegates, and in the early 1840s, he presided over that council with an income of twenty-five thousand francs (AP, Archives départementales de la Martinique, *procès-verbaux du Conseil colonial*, 10 February 1841). In 1842, French sugar production had declined sharply because the resumption of cane imports from the colonies undercut prices, but by 1842, France had sixty sugar beet factories producing two pounds of sugar per capita annually. Partisans of the colonialists, Dupin among them, won a minor victory in 1843 when the Chamber adopted a law in favour of progressive taxation on home-produced sugar beets to equal the tax that was imposed on imported sugar from the colonies. However, this came too late for the colonialists; the entire question was linked to the move for the abolition of the slave trade, to which Dupin was opposed. Was this the Dupin who claimed to foster the interests of the *petit producteur français* (French small producer) and French industry?

The list of his speeches continues. On 3 July 1843, he reported on the examination of a bill for the opening of an exceptional credit of 5,490,000 francs for the expenses of French establishments in Oceania.[4] On 12 May 1845, he reported on a bill for a sum of 112,000 francs for the acquisition of the mineral collection of the Marquis de Drée.[5]

On 17 April 1846, the Chambre des députés had decreed the allocation of 93 million francs for the navy for construction and arsenals. Lamartine had spoken on this subject on 15 April, expressing his fear of the power of England. On 15 June, Dupin submitted a report to the Chambre des pairs in favour of the bill, which was passed by 122 votes for and two against. Dupin sent a copy of his report to Lamartine with the following comment:

> I am sending you my report to the *Chambre des pairs* on the decree of 93 million for the navy. The success of your admirable speech in the *Chambre des députés* followed us into the other Chamber, where we were able to achieve unanimity minus two … You brought me luck, or rather, your support in favour of my work predisposed universal approval (M.-R. Morin 1995, 26)

On 20 February 1847, he reported on a bill concerning the coastal trading of grain by foreign ships. In addition to his work in the Chambre des pairs, Dupin was again elected, on 10 September 1843, the mayor of Saizy (see chapter 1).

17.5. Dupin, Member of the Assemblée Nationale Constituante (1848–May 1849)

Returning to the National Assembly in 1848 as a representative for the Seine Inférieure, Dupin contributed to discussions on pensions for naval workers (on 22 September 1848), the organization of the Admiralty Council (on 15 November 1848), and the naval budget for 1849 (on 14 April 1849). Always concerned with the navy, he had already reported four times on the budget for that service prior to his brief ministry in 1834 (see chapter 17.3). In July 1849, he spoke on the abolition of poverty; on 8 August 1849, on the work of *ponts et chaussées* engineers; on 18 February 1850, on Algerian trade with France and abroad; on 6 July 1850, on the expenses related to the Great London Exhibition; and in 1851, on the salaries for cardinals. This gives an indication of the diversity of his contributions.

17.6. Working Conditions and Child Labour

It must be said that despite the contradictions in his character and the accusations of paternalism in his attitude to the working class, Dupin did constantly strive to better working conditions. Since his time in Corfu, when he had been concerned with the lot of military workers (JEP, 1815, vol. 10, *cahier 17)*, he had favoured the social advancement of the most severely disadvantaged people. He maintained that industrialization would not result in the impoverishment of the working class and that the development of technology in production could only better their conditions. Workers should live in harmony with their employers, he argued, and have recourse to them to resolve temporary crises in industry. The education of the workers, which he so constantly championed, would improve their lot so that in their turn, they could become employers in the small businesses which he considered to be the basis of French wealth.

Thus, Dupin felt, all would be well if the new industries would not make use of child labour. He considered that children should be protected as the future of France, and he applauded the British 1833 Factory Act, which outlawed the employment of children under the age of nine in the textile industries, limited the working hours of women and children, and instituted regular factory inspections. In 1840 in the Chambre des pairs, a commission was appointed to study a bill on child labour in industry, with Dupin as its reporter (see Anceau 2009, 253–270, for details of the work of the commission). He said,

> What do we want from this law? To establish, for children, protective conditions, which, in no instance, may be violated. What are we giving to the Government? The increased power to protect children; the facility even to reduce the hours in the work we are establishing, if, for some industries, it considers it overwhelming and harmful. (Dupin 1840b, 15–16)

Was his attitude paternalistic? He was a man of his time; for him, the improvement of the lot of the workers and children went hand in hand with industrial development and a thriving economy. In France, the age of admission to factory work was generally six years, but an enquiry of 1837 had revealed that children as young as four were being used for simple operations in the textile industry. Doctors and economists, disturbed by this situation, sounded the alarm (see Léon 1968):

> Regular use of certain movements caused physical abnormalities such that these children, through pushing the bobbins, became deformed and, eventually, lame. Still worse, the children worked at night, where their assistance was essential to the adults, and there was no protection for them. (Fohlen 1967,)

Debates on the subject continued through 1840 and 1841. The speech made by Dupin, the reporter on the bill, indicated that 'of ten thousand conscripts from the industrial départements, 8,980 were disabled, and the proportion of those who were unfit for military service was 4,029 from the agricultural regions' (Pernoud 1960, 2:392). Finally, a law was passed in 1842 forbidding the employment of children under the age of eight and night work for children younger than twelve.

Dupin did not stop there. On 15 February 1848, just before the Revolution, he presented his *Discussion sur le travail des enfants, des adolescents, des filles et des femmes* (On the work of children, adolescents, girls and women) to the Chambre des pairs. This was a cause with which he had long been involved (see Dupin 1840b). He sent a copy of his speech to the Chambre des pairs to the Duc de Montpensier on 3 July 1847 with the comment,

> Allow me to offer you my report on the work of children in factories and workshops of the kingdom. The future of the people is in the good we wish to do through the education of children, whose strength we spare through limiting their work. (ADN; PJ 89)

Apart from his work in Parliament, Dupin continued to interest himself in the welfare of children. Perrin (1983, 666) referred to a public meeting of the *Association des fabricants et des artisans pour le placement et l'instruction des jeunes orphelins* (Association of industrialists and artisans for the placing and education of young orphans) on 2 July 1848. There, Dupin spoke at some length and stated,

> This association for helping the children of the people, it is the people themselves who constitute it and support it through voluntary sacrifice. The members of the association are, firstly, good

> directors of workshops and factories, and generous artisans who open their workshops to receive our orphans as apprentices, whom they treat as equals. (Perrin 1983, 666)

Dupin's work on behalf of children was bearing fruit. At the conservatoire, in the press, and in Parliament, he continued to try to further the interests of the working class, as may be seen in some of his publications of this period (Dupin 1831b, 1833a, 1841c).

17.7. THE *CAISSES D'ÉPARGNE* (SAVINGS BANKS)

Whatever else Dupin may have achieved during his political career, he was particularly successful in the establishment of savings banks, providing financial protection for working people (see Christen 2009a, 235–252). Their creation took place in the context of the social concerns that were developing from 1820 through 1840. Dupin had been impressed by the system of savings he had seen when he visited the works of Alexander Galloway, one of the most important machine manufacturers in London. Moreover, Galloway insisted that his workers should be able to read and write, and he stressed that the conduct and lifestyle of his workers had greatly improved during his thirty years of implementing this policy (Dupin 1825c, 182).

In Galloway's factory, Dupin explained:

> 12 sous a week are taken from the wages of every worker. As soon as one falls ill, he receives from this fund 25 francs a week. The ordinary labourers pay only half … Every year at Christmas, … if receipt exceeds expenditure, the surplus is divided among all the workers, according to the proportion of their investment. (Dupin 1825c, 184)

In his *Petit producteur français*, Dupin stressed the advantages to small businesses, craftsmen, and humble workers of depositing savings. In 1831, he became the administrator of the Paris *caisse d'épargne*. On 16 May 1834, he insisted that the administrators should be appointed from among government representatives, conseils généraux, and the municipalities as well as ordinary citizens.

On 22 March 1837, Dupin addressed the conservatoire on the subject, explaining that savings banks had begun in Britain in 1798. At first, in opulent England, they achieved little success, but in Scotland, a poor country, 'austere and careful', they had developed rapidly (Dupin 1837a). In the same year as the first savings bank opened in London, one was founded in Geneva under government control. In 1818, the first French *caisse d'épargne* opened in Paris. Eight years after the establishment of the Paris bank, in 1826, workers represented merely a sixth of the depositors. However, the proportion rapidly increased; in 1827 and 1828, it reached 30 percent of new depositors, then 40 percent in 1830, 45 percent in 1831 and 1832, and 52 percent in 1836. Dupin stated (1837a):

Today, the number of depositors has reached a total of 82,709, approximately distributed thus:

workers, male and female 43,000
servants 17,000
others......................................22,609
Sums deposited ....48,834,763 F
Average per person ....590 F 88 c

He then went on to describe the situation in the provinces, where the *caisses* had developed later and had made slow progress until 1837, when there had been a prodigious increase in deposits. He continued in the elaborate terms of a politician:

> The *caisses d'épargne* represent the patent and undeniable proof of the prosperity, welfare and progressive security of the French people; they respond victoriously to the sophistry of those who claim that the constitution of societies produced by modern civilization, is incapable of insuring the lives of the working-class. Such an attitude is the source of the hatred shown to the institution of the *caisses d'épargne* by men whose ideals rest on the hypotheses of oppression, misery and the starving poverty of the working-classes, on whom they call to reverse the present system. (1837a)

Dupin's verbosity had been severely criticized in 1838: 'He is quite incapable of restraining the flow of his inappropriate eloquence. He has to talk, talk, talk' (Timon[6] 1837, 199).

In 1840, he again addressed the conservatoire on the *caisses d'épargne*, this time in his opening speech to the course in geometry applied to arts and statistics:

> May we be permitted to congratulate ourselves on this progress, to which our feeble efforts, our lessons, our calculations have, perhaps, contributed, bringing confidence and conviction to the people of Paris. Let us now look further to the entire kingdom; since the law on the *caisses d'épargne*, in which we acted three times as reporter, the number of *caisses* has increased, in six years, from thirty to more than 200, established in all regions of the country. (Dupin 1840b, 50)

Three years later, Lamartine, in the journal he had recently founded, *Le bien public*, commented,

> 'M. le baron Charles Dupin has just published a luminous and eloquent defence in figures of this institution [the *caisses d'épargne*] which already has 450 deposit banks in our departments, and invests 50 million a year in the reserves of the people. It is essential reading. This man has really invented a new language, particular to algebra: the language of investigation. Figures were just numbers, he makes of them reasons.' (1st year, no. 4, December 1843)

Lamartine had taken an early interest in the *caisses d'épargne*. A champion of the antislavery movement, he must have had serious differences of opinion with Dupin, but on this subject, they concurred. On 6 July 1834, Dupin wrote to Lamartine, who was the député for Mâcon:

> Allow me to request your support to obtain from the *Conseil général* of the Saône et Loire, in the next session, the bill on the *caisses d'épargne*,[7] a bill which I believe I have improved by various clauses to make it more completely advantageous to the working classes. This resolution will be of the greatest use to us in overcoming the difficulties raised by the Finance administration. (M.-R. Morin 1995, 9)

On 11 July 1834, Lamartine replied expressing his support but also noting that 'this element in the improvement for the people, the *caisse d'épargne* should be but a seed for more philanthropic institutions' (M.-R. Morin 1995, 10). When Dupin was appointed to the peerage, Lamartine wrote to him on 7 November 1837, 'No-one in the Chamber was more delighted than I by your promotion to the peerage' (11). He continued in his friendship and support of Dupin. When on 8 May 1842, Dupin addressed the conservatoire on the *Progrès moraux de la population parisienne depuis l'établissement de la caisse d'épargne* (Moral progress of the people of Paris since the establishment of savings banks)—which was subsequently reedited and incorporated in his 1844 *Constitution, histoire et avenir des caisses d'épargne de France* (Constitution, history and future of the savings banks; Dupin 1844d)—Lamartine wrote to him on 14 December 1842,

> Dear and eloquent colleague,
> I have received, read and applauded in my heart even more than with my hands. That is what you inspire in the people who admire you. They feel that the prize you seek from your work is not a prize of vanity but of the public good. (M.-R. Morin 1995, 22)

Lamartine had probably read Dupin's *Progrès moraux de la population parisienne* (Dupin 1842b). Dupin would reproduce it in 1844 when he grouped together his main reports and speeches on the *caisses d'épargne*, including his addresses to the conservatoire, between 1834 and 1843. In this work, *Constitution, histoire et avenir des caisses d'épargne de France* (Dupin 1844d), he recalled the important contribution of the first presidents of the *caisses*, the Duc de la Rochefoucauld-Liancourt (who held the position from 1818 to 1827) and his successor, Baron Benjamin Delessert. Dupin insisted that the administrators should be appointed from among government representatives, the conseils généraux, and the municipalities as well as from ordinary people. He referred to the many workers who had achieved success in their professions and to the female domestic servants who had made use of the *caisses d'épargne* to their advantage—indeed, using the opportunity to 'fill their bottom drawer' before returning to their villages to marry (Dupin 1844d, 282). He observed that many of the women saved for many years to marry, somewhat late in life, younger husbands. He argued for the great advantage of the banks for women, noting that prudence and order in the home depended on them 'much more amongst the working-classes than in the higher ranks' and that they kept a firm hand on the family purse strings. Savings also gave people access to the possession of property. Dupin the economist revealed himself again:

> It is an illusion to think that the *Caisses d'épargne*, when they receive the workers' savings, take essential capital out of circulation … they return the monetary wealth of France; they cause it to flow over the entire territory through the activity of 600,000 depositors who, on the one hand, invest an annual 130 million in deposits and, on the other hand, withdraw 80 million to satisfy, with double the efficacity, the alternative needs of circulation. (1844d, xxxviii)

However, Dupin's arguments failed to convince the opposition, and on 22 June 1845, a law was passed to lower by half the total investments in the *caisses d'épargne*. A decline began in the banks. In 1848, there was widespread revolutionary turmoil in Europe, and in the face of failed harvests and a financial crisis, France was once again in a state of revolution. During a discussion in the Chambre des députés on the compulsory conversion of the *caisse d'épargne* accounts at eighty francs—as opposed to one hundred francs, their normal value—at an interest rate of 5 percent and subject to stock exchange fluctuations, Dupin was obliged to defend the banks on behalf of the depositors, who were likely to become the victims of speculation. On 8 July 1848, the value of the deposits fell to seventy francs, meaning that those people who needed access to their deposited money stood to lose a considerable sum. After a sudden rise in the deposits' value, Dupin declared,

> We need to know if we are transforming the *Assemblée nationale* into a gambling den. It is 4 o' clock, and they want to debate,

> what? The stock exchange rate as it was quoted just an hour ago. I protest against this type of action. What, gentlemen! You do not know if such a sudden and rapid rise over 24 hours is not the result of scheming. You have no proof; and when it is a matter of gaining advantage from this extraordinary increase, to the detriment of whom? 800,000 families of which 600,000 belong to the working-class for whom your consideration is constantly invoked, and you would use against these families a rate that has only just been quoted. To me, that is unacceptable. I declare that never, in any question of finance, never has been introduced, in the increasing of any type of fund, the level of the stock exchange at the moment of the vote. I protest, in the interests of the people, against this harsh move. (AP, *Comptes rendus des séances de l'Assemblée nationale constituante du 4 mai 1848 au 27 mai 1849*, 3 July 1848, 2:384)

This was one of Dupin's last speeches in the Assemblée constituante. The year 1848 saw the abdication of Louis Philippe, an end of the Bourgeois monarchy, and the creation of the Second Republic, which was followed in 1852 by the Second Empire. During the Second Empire, probably deeply disappointed by the decline of the *caisses d'épargne*, Dupin said no more in the Chamber on the subject. In his 1842 *Progrès moraux de la population parisienne depuis l'établissement de la caisse d'épargne*, Dupin had pointed out that Great Britain had been far more reasonable than France in declaring that

> any individual depositing without interruption just 6 francs a month in the *caisse d'épargne*, starting before the age of 30 and continuing till 60, would receive an income of 500 francs for the rest of his life once he had reached the age of sixty. (1842b)

On 6 June 1851, he wrote dispiritedly to his friend Lamartine, 'The *caisse d'épargne*, savings bank in English, is the bank of the people … In England (a monarchy), royalty permits a man of the people to have as much as 5,000 francs on deposit in the people's bank' (Morin 1995, 27).

In France in 1834, the constitutional monarchy had permitted an investment of three thousand francs; this was reduced to two thousand francs in 1846, and a further reduction to one thousand francs was proposed in 1851. Dupin called on Lamartine: 'Attack this infamous proposal which, to humour the bankers of the Pactolus,[8] reduces to the infinitely small the bank of the little people' (Morin 1995, 27–28).

17.8. Opposition to the Socialists

Dupin was to make another important speech in August 1848 in a discussion of the plan to abrogate the decree of the previous 2 March concerning the limitation of working hours. The debate was heated, and Pierre Leroux (1791–1871), a publisher, philosopher, and politician who had recently been elected the representative of the Seine department as a democratic socialist, declared, 'Let the captains of industry who encourage or demand a 14 hour day, not come and claim that their workers consent to this' (MU, 188, 6 August 1848, 1576).

Dupin leapt to the defence of the industrialists, claiming that industry was not responsible for poverty and that the difficulties resulted temporarily from commercial crises, but mainly 'from socialist theories planting error and hatred in the hearts of the workers' (see Anceau 2009, 263). Using statistics to support his claims, he showed that the decree of 2 March had been detrimental to industry and therefore to the workers themselves, producing unemployment and limiting the purchasing power of the labourers. He was in favour of increasing the working hours to fifteen and added, 'This maximum, we introduced in January 1848, in the law on the work children in factories. If it was judged good for women and adolescents, it must also apply to men' (MU, 245, 1 September 1848, 2248–2250).

He continued to battle with the socialists, and on 15 November 1848, in a debate on workers' associations, he declared that they would be the ruin of family workshops and small businesses. At the conservatoire on 17 December of the same year, he vehemently attacked the socialists for having 'in ten months, ruined France for ten years' (Dupin 1848, 61–71). He was opposed to the 1851 coup d'état, but he protested only desultorily against it before accepting a seat in the Senate in January 1852.

17.9. Dupin, Senator of the Second Empire (1852–1870)

In the Senate, where he was referred to as 'M. le baron Charles Dupin', there was a notable variety in the topics on which he chose to speak—for example, the organization of the Institut (on 9 April 1862), the preservation of forests (on 27 May 1865), public works in Algeria (on 7 July 1865), public libraries in Saint Etienne (on 25 June 1867), crèches and the related progress of industry and the population (on 7 January 1870), and the causes of depopulation (on 11 January 1870). This is a brief summary of all of his various contributions.

According to Anceau, 'in this Assembly [the Senate] Dupin showed himself at ease and to his advantage. He was one of the most active senators and spoke in most of the debates' (2009, 268). There were major changes in the 1860s, when Napoleon III found himself abandoned by the gentry, clerics, and protectionists and had to find support among the lower classes. Dupin saw this policy as dangerous and a way to allow socialism to rear its head again. A plan to change articles in the penal code so that workers' associations would be permitted was symptomatic of this new political development. He and his elder brother opposed the move, but they were in the minority, and on 25 May 1864, 'the law granted the workers what all governments had refused them since 1791' (267). Dupin bided his time, and when the manufacturers of luxury furniture presented a petition against a strike for higher wages among their workers in 1865, he was able to criticize the new law. He stressed the negative aspects of strikes, not only for the manufacturers but for the workers themselves. He made little headway.

The question of child labour was again raised in 1864. Reports described the appalling conditions of the children in the manufacturing industries; at the end of a long and arduous day, they were in intolerable physical condition and were quite unable to study. Dupin called once again for the creation of an inspectorate for the defence of children, observing that the existing controls by préfets and voluntary workers were inadequate. His argument was convincing, and on 7 December 1868, an imperial decree reorganized the system of inspection, entrusting it to mining engineers, who were already experienced in applying the law forbidding children under the age of ten to work in the mines (MU 344, 9 December 1868, 1551). Dupin had a humane approach on certain social issues—child and female labour, for example. However, he did not seek to put an end to this practice but rather to prevent its abuse by industrialists. He was totally opposed to socialism and fundamental social reforms. It became increasingly apparent that his main concern was for the welfare of industry, with workers' education directed to that end; for him, the wealth of the nation was paramount.

I shall now look at his speech to the Senate of 25 February 1870 (AP, Archives du Sénat, 1870) on 'the freedom of higher education and the decentralization of the University', a subject that was close to his heart. The universities had been closed during the Revolution; the Ecole polytechnique had been established, and the existing *grandes écoles* (top schools), such as the Ecole des mines and the Ecole des ponts et chaussées, were reorganized. In 1806, Napoleon created the Université impériale, a national organization incorporating all secondary and higher education, divided into regional academies; this was to become the Ministère de l'instruction publique in 1828. Moves began around 1870 for the creation of real universities for higher studies, but there was opposition to this proposal because autonomous universities were seen as a threat to the state organization of education. It was not until 1896 that modern universities were finally created in France. Dupin supported the cause of autonomous universities. In discussing the subject, he drew on his knowledge of Britain, speaking of the universities of Oxford, Cambridge, Saint Andrews, Glasgow, and Aberdeen, all of which were independent of the government, before referring to Belgian, German, and Italian universities. Of Paris, he said,

> Remember that this ancient and glorious University of Paris, which preceded all others on the continent, which had, from its inception and its organization, rendered infinite service since the Middle Ages … Professors from all over Europe came to study in Paris … to enhance their knowledge … before returning to their own countries … to copy the French organization. Could

> it be claimed that there were 23 superfluous universities and that a single one would have sufficed? They represented the intellectual attainment and honour of our ancient provinces which had been nations, and preserved, as regards the culture of minds, this noble fruit of their partial nationality … These universities were undeniably necessary to their respective people, to their respective provinces … In 1789, they were abolished by one stroke of the pen … Now, … a commission is appointed to examine the serious question of higher education. This commission will see if it is absolutely necessary that there be one single university or if it would not be more advantageous, even better, to restore several of them. (AP, Sénat, 25 February 1870)

This 1870 speech was one of the last of Dupin's last political contributions. His political activities had spanned the different regimes of France from the Restoration to the advent of the Third Republic. He was now eighty-six years old and would die three years later.

ENDNOTES

1. Antoine, Comte de Argout (1782–1858), a member of the moderate right and pair de France.
2. The identity of this person has not been determined. As the letter is in the Nièvre archives, it may be that he was a local acquaintance.
3. Etienne Maurice Gérard, Comte de Gérard (1773–1852).
4. Oceania, the region between Southeast Asia and South America, comprising Australia, New Zealand, Papua New Guinea, East Timor, and other islands and archipelagos.
5. Marquis Etienne de Drée (1760–1848), a French mineralogist, geologist, and agronomist. Apart from his own collection of minerals, he had preserved that of Dieudonné Dolomieu, also known as Déodat de Dolomieu (1750–1801), the famous geologist from whom the rock dolomite derives its name.
6. 'Timon' was the pseudonym of Louis Marie de Lahaye Cormenin (1788–1868), a left-wing parliamentarian.
7. The bill would be discussed in the Chamber in February 1835.
8. The legendary river, a source of great wealth in Midas's ancient kingdom of Lydia.

Chapter 18

Dupin's Naval Career and His Last Contributions

18.1. The Early Years With the Navy

To begin summarizing all of the varied activities of Charles Dupin, I shall start with his naval career. It began on 4 November 1801 and officially ended on 3 April 1852—a total length of service of fifty years and five months. His active service actually ended on 1 September 1818 when the minister for the navy and colonies granted him residence in Paris. By then, as has been seen (in chapters 4 and 5), Dupin had pursued a brilliant career in Dunkirk, Boulogne, Toulon, and Genoa, followed by his first two research visits to Great Britain when he was an ingénieur de 3ème classe. From September 1818 on, he served in an administrative capacity, but he would attain the rank of inspecteur général du Génie maritime. At the same time as he was involved in teaching, economics, and politics, he climbed the higher rungs of the ladder in his naval career between 1819 and 1852. The details of that career are given in a document in the naval archives (ASHM CC7 779) and signed 'Charles Dupin, member of the Academy of sciences'. His report is as follows:

1. 18 years of complete service.
2. 5 war campaigns.
3. 12 years commanding military workers: more than 600 workers trained by M. Dupin in naval construction, troop exercises and fortification works.
4. Sailing with the squadron of Admiral Ganteaume in the Mediterranean (1807).

During this time with Ganteaume, the admiral's three-deck vessel *La ville de Paris* lost all of her sails and topmasts and her lower yardarms were broken. Admiral Ganteaume (see chapter 5.1) reached Corfu, entrusted to Dupin his flagship with his military workers, and left with half of his squadron to go in search of the Cosmao[1] division, according to Dupin's account. When he returned a week later, he found the upper masting completely restored and the sails hoisted. He boarded his ship, reached the English fleet, and led his squadron back to France with complete success.

5. Port service: Involved in the work for the creation of the flotillas of Antwerp and Genoa.
   Introduced in France iron cables, ropemaking processes, the English system of oblique framework, recently adopted for the frigate of M. Willaumez [L. E. B. Willaumez (1808–1871), French admiral].
6. Voyages to Belgium, Holland, Italy, the Ionian Islands, England, Scotland and Ireland: crossing and recrossing the British seas from south to north during the most varied seasons. A leg dislocated in Ireland.
7. Scientific work related to the navy:
   Applications of geometry and mechanics to the stability of ships, strength of timber and the resistance of the framework of ships, etc.
   Tableau de l'architecture navale aux 18e et 19e siècles.

18.2. 1834 to 1843 (Under the July Monarchy)

On 10 May 1834, it was proposed in a report that Dupin should have the grade of director of naval construction, but he did not achieve this until 1838 (ASHM CC7 779, Service du personnel du ministère de la marine). On 16 June 1834, Dupin's salary was fixed as a delegate for Martinique (see chapter 17.4).

In a report to the king on 18 November 1834, the minister for the navy stated that Dupin had sat on the admiralty council from 25 July 1831 to 10 November 1834. His reinstatement was requested and granted starting on 18 November; the interval between 10 and 18 November corresponded to Dupin's brief service as minister for the navy, when his holding the two positions would have been incompatible.

When Dupin received his promotion to director of naval construction on 15 May 1838, his father wrote him, for once, a kindly letter:

> Your brother, my friend, rushed cordially to give me the official news of your new appointment to the higher grade of Director of naval construction, but my satisfaction with regard to this is not unreserved, because of my fear that the accumulation of so many different duties may compromise your health.
>
> In the past, I had to push my young sons hard on the wheel of work but, at their present age, I must now exhort them to 'slow down' so that they are not dragged to an early grave. (DRB; PJ 90)

On 19 July 1843, Dupin was appointed inspecteur général du Génie maritime. This, like membership in the Admiralty Council and the directorship of naval construction, was purely an honorary post. However, when the National Assembly was dissolved in 1851, Dupin returned to the aegis of the navy and to his old post, inspector general of the Génie maritime. On 14 January 1852, a salary was requested for him by the naval authorities (ASHM CC7 779). On 3 April 1852, Louis Napoleon, president of the French Republic, permitted Dupin to justify his claim for a pension; it was granted, and Dupin was awarded an annual 4,800 francs beginning 1 August 1853. The sum was calculated from the time he had entered the Ecole polytechnique in 1803 until 7 June 1853, but it was a considerable drop in income. Where previously he had had 17,000 francs from the navy, 5,000 as professor at the conservatoire, and 1,500 as a member of the academy—a total of 23,500 francs—he now had only 4,800 francs plus the income from his work at the conservatoire and his membership in the academy, which came to just 11,300 francs. The next note on his record concerns the settlement of a pension on his widow. For a detailed account of Dupin's naval career, see Junges (2009, 43–52).

18.3. SAINT-SIMONIAN, FREEMASON, SPY?

The Saint-Simonians were in the vanguard of industrial development in France and exercised considerable influence within the Ecole polytechnique; indeed, many former *polytechniciens* were numbered in their ranks. However, no evidence has been found that Dupin was one of them; he did not contribute to their journals. As for freemasonry, according to records in the Nièvre archives (ADN file 4 J), both of Dupin's brothers belonged to the order, as did Monge and other eminent engineers and scientists. Perrin (1983, 693), who made an exhaustive search of the records, maintained that there is no indication that Dupin was a member.

As for espionage, it seems to be tempting to look for evidence of this in history. It has been seen (in chapter 10.2) that Dupin was a great patriot, and undoubtedly, he used his time on his travels to observe everything that he believed could be of benefit to France. Unlike the many industrial spies described by Harris (1998), who came to Britain in the interests of their own industries with the goal of copying as much as possible, or military visitors such as Forfait and Lescallier (see Bradley 2005), who visited shortly before France and Britain were at war, Dupin's visits were made in a time of peace and were very much intended to enhance his career. He learned a great deal, not least of which was the value to industry of educating workers. He took careful note of all he saw, but he visited workshops and naval installations with an open curiosity.

His reports were certainly useful to France, but can this be described as espionage? Galloway (chapter 10.8) certainly suspected Dupin's motives. Perhaps this subject should be left open to question; perhaps Charles Dupin was simply what he appeared to be.

18.4. The Declining Years

The outstanding mathematician and naval engineer were very much a part of the past during Dupin's last years. As has been seen, he still held posts in the navy until he retired from the service in 1853, but they were administrative and honorary; by that time, he had spent more of his life in politics and political economy. He did not retire from the conservatoire until 1864, but it is probable that his assistant Aimé Laussedat, a member of the Institut, played an increasingly important role in teaching and substituting for Dupin. Throughout the Second Empire, seeking more often the peace and quiet of his native countryside, the ageing Dupin was preoccupied with general questions of politics, economics, and local matters.

On economic questions, he took a particular interest in India and wrote extensively on the subject in his *Force productive des nations* (Dupin 1867b). He was fulsome in his praise of the British India Association, whose foundation in 1851 was a major event in nineteenth-century Indian history. The association's aim was to secure improvements in the administration of the country as well as reform and restructuring. Dupin commented,

> Never would the Romans or the Spaniards have permitted that, within their recent conquests, independent associations be constituted with such an aim; but they are typical of this noble sentiment of freedom which characterizes England and which she takes everywhere, as a principle of enlightenment and greatness. (1867b, iv)

Apart from his elaborate writing style, Dupin provided much in the way of analysis and statistics in this study. When he sent his manuscript to an unnamed general, he wrote in an accompanying letter on 21 July 1862,

> In this introduction, I wanted to summarize, quickly and in general terms, the characteristic interests, the extent of powers and the special accomplishments of the nations I have described … I have already written on the productive strengths of 900 million people; I have just 450 million to cover … this is the main and last concern of my life. (ADN; PJ 90)

At the age of seventy-eight he was now, in his declining years, spending most of his time in the Château de Réconfort, his property in the Nièvre, and was writing from there. He was seventy-eight years old. Four years later, in 1866, he wrote to the historian Pierre Nicolas Rapetti (1811–1885), who was responsible for the publication of Napoleon I's correspondence,

> Château du Réconfort, par Monceaux-le Comte, 8 September 1866
>
> I am continuing with my work on that part of the Orient which is not included in the India question; I am at the Introduction of my last part and I have some magnificent [illegible, perhaps 'details'?], I would not say to develop but to condense. (ADN; PJ 91)

His handwriting in this letter is that of an old man.

Some time later, he wrote again to Rapetti (ADN n.d.; PJ 91) that the printing of the last page of the seventh volume of the *Force des nations* was finished; he was busy correcting the eighth volume. He had left the India question and was concentrating on Islam, explaining that his strength was failing only gradually and that his patience was intact. The work appeared in 1867 in seven volumes—one each on the United Kingdom, America, the Orient, Asia, Indo-China, and India, with the two final volumes devoted entirely to India. At Dupin's funeral, General Morin referred to the work: 'The book will not be finished, but the completed parts are imposing' (A. Morin 1873, 21). It is to be wondered how, with no firsthand experience in other countries apart from his travels in the United Kingdom, Dupin could devote himself to such a work and where he acquired all of the necessary information and statistics.

The Second Empire had come to an end in 1870 with the proclamation of the Third Republic. Defeated in the Franco-Prussian war, France had lost its eastern territories of Alsace-Lorraine, but resistance continued under the government of national defence. The republican government, which was at first amenable to the terms of the armistice, renewed the declaration of war, but Léon Gambetta's provincial armies suffered continual losses. The siege of Paris from September 1870 to January 1871 brought about the final defeat of the French. Paris was starving, and in February 1871, it was obliged to capitulate. The German armies entered the city in triumph. Still defiant, Parisians armed themselves, demanding self-government, and the Paris Commune was created on 28 March; one of its leaders was Louis Auguste Blanqui, a brother of Dupin's economist colleague at the conservatoire (see chapter 15.4). This was the first example, probably in the world, of government by the working class, and other cities, such as Lyon and Grenoble, followed suit. After a rule of two months, the Commune was crushed and eliminated in May 1871, but in July, elections strengthened the left wing in the Assembly.

Gambetta (1838–1882), the founding father of the Third Republic, was the minister of the interior during the government of national defence. He championed the freedom of the press and universal suffrage. Elected as a deputy for the Seine department in 1871, he defended the Republic against the restoration of the monarchy. People like him did not at all suit the archpatriot Dupin, in whom some of the old fire remained, as can be seen in his comments to Rapetti on the political situation in a letter dated 1 September 1871:

> You may understand with what patriotic and grievous anxiety I study the phases of the proposals and deliberations on the functioning of the constituent government and the Presidency … What will be the future, too close, perhaps, of this unprecedented system … I found Gambetta inept, without skill and violent. Let us be careful not to complain about him. (ADN; PJ 92)

This was another turbulent period for France, but Dupin, in his old age, was devoting more and more time to his work as the mayor of Saizy, local affairs, and his family. However, as seen in the letter he wrote to Pasteur in 1872 (chapter 1 and PJ 4), he was still keeping abreast of scientific developments. He stayed either in the Château du Réconfort—whence he had written, sixteen months before his death, 'And more than ever, in my solitude, I breathe in the peace and independence which must be the happiness of the wise man' (letter of 1 September 1871 to Rapetti (ADN; PJ 92)—or in his Paris home in the rue du Bac, where he died on 18 January 1873.

18.5. Death and Eulogies

At Dupin's funeral on 21 January 1873, General Arthur Morin spoke on behalf of the Academy of Sciences and the conservatoire. He described Dupin as a great patriot and summarized his contributions and his life's work. He stressed Dupin's courage in praising Monge as 'one of most undeniable glories of the *Institut*' (Morin 1873) when Monge had died in political disgrace. He spoke of Dupin's love for the beauties of geometry and his conviction of the importance of observation and experience in engineering. Above all, he dwelt on Dupin's efforts for public education and quoted extensively from his opening speech to the conservatoire on 30 November 1826, adding that 'for him, public instruction was essentially a productive force' (A. Morin 1873, 19).

Joseph Bertrand (1822–1900), the celebrated mathematician and historian of sciences, said of Dupin,

> 'Continuing his teaching at the Conservatoire, untiring at our sessions and those of the ASMP [Académie des sciences morales et politiques], he carried out to the letter, until the age of 88, his double duties as academician; moreover, member of the Admiralty Council, delegate for the colonies, so much study and work became, on many occasions, for the *député*, the *pair de France* or the senator, a preparation and an aid.' (Bertrand 1888, xxiv)

Charles de Lévêque, a member of the Académie des sciences morales et politiques, declared in his funeral oration for Dupin,

> The totality of his works represents a sort of treatise on the relationship of the physical and moral qualities of the people. In his hands, statistics, this science of numbers, comes to life, inspires, becomes clear ... through numbers, he aimed at the improvement of customs. He studied the science firstly for itself then, immediately, to apply it. He had faith in the virtue of science for the common good. He used it to raise up the labouring classes. (1873, 99:609)

In the twentieth century, Fox observed, 'Dupin's talents and the long and distinguished public life in which he applied them were dazzling' (1985, 304). Fox's description of the young, enthusiastic, and talented young engineer's arrival in Corfu has been discussed (chapter 5.1). He stressed Dupin's conviction that education, with an emphasis on mathematics and science, was the key to prosperity and social progress.

Gueslin described Dupin as one of those important characters in history 'who have left no very visible traces in historical records' (2009, 271). However, the more obscure historical players can also have a decisive influence, both in their lifetime and the future, and in his time, Dupin's scientific, economic, and social writings were of great significance. According to Gueslin (2009, 275), it is for his work on social issues that Dupin is well-known—his work for women and children in the manufacturing industries, for example, and his advocacy for workers' education and the *caisses d'épargne*. It is time for the full importance of his contributions to be recognized.

18.6. Conclusion: A Man of Contradictions and Many Facets

Who and what was the real Charles Dupin? He cannot be placed in any particular category. He began his career as a brilliant mathematician and would certainly have shone as a naval engineer if he had continued with that work. He became interested in social issues and travel; Britain,

France's archrival in industrial progress, was soon his focus of study. He produced a remarkable record of his six visits to the British Isles. In his writing on the subject, he not only observed and noted in meticulous detail all that he saw of industry and military prowess, but it was there that his expressed determination to support the education of the working classes began.

As an educator, he had a fine career; as an economist, his attitudes varied. As a politician, he was first opposed to colonization, but then, as has been seen, he became a fervent partisan on the issue. It is hard to reconcile his concern for the working classes with his defence of the slave trade, and he was accused of being completely ignorant of the needs and conditions of the workers (chapter 16.6). However, he was man of courage and defended his causes rigorously.

Dupin was such a multifaceted character that he failed to reach the heights he could have achieved. As Fox put it,

> It may well be that Dupin's reputation suffered in Tannery's day [see Tannery 1887; he had little time for Dupin] and has suffered ever since from the very diversity of his activities and the success he enjoyed in all of them … this intellectual adventurousness drew in important ways on a creative interaction between his observations abroad and the turbulent economic and social history of his own country. (1985, 319)

About Dupin's travelogues, Fox further commented that they were 'documents of outstanding importance for the historian'.

This has been the story of a great man manqué. With his knowledge of ship design, coupled with his appreciation of the arts and his love of language and literature, Dupin might have rivalled Sané in the creation of beautiful ships. He wrote excellent records of his travels, full of interesting details, and his powers of observation were remarkable. As a mathematician, he showed extraordinary promise when he was a young *polytechnicien*, and he could have taken his work in this area much further. Instead, he charmed his way into all sorts of social circles and enjoyed friendships with and the appreciation of the great people of his time. In his travels, he easily made contacts with people who obviously enjoyed his company and were eager to see him again. He even charmed the dour and wary James Watt.

The seeds of social ambition were sown early in him, and he explored all possibilities to ensure his advancement rather than concentrating on his naval career and mathematics, in which he would have excelled. Dupin's activities in providing courses for the working class changed the face of education in France. However, writing in the *Grande encyclopédie*, Paul Tannery (1887–1902, 81) dismissed him peremptorily as an opportunist and a careerist. He was certainly determined to get to the top personally as well as to see France achieve great things. Very much a patriot, Dupin's concern for workers' education was largely prompted by his desire for France to develop its industry and rival the Britain he so much admired. Nevertheless, he left his mark in that realm of education and many other areas; mathematicians are well aware of his *cyclides*, and statisticians quote his work. He was very much a man of his time, one of the great characters of nineteenth-century France—and as such, he was an outstanding figure who deserves his place in history.

ENDNOTE

1. J.-M. Cosmao-Kerjulien (1761–1825) was greatly respected by Napoleon.

Pièces Justificatives

All documents, letters and extracts are reproduced *verbatim.*

1. From Dupin's letter of candidature for the Academy of sciences (6 September 1818), (AAS):
   J'ose me mettre sur les rangs pour la place éventuellement vacante dans la section de mécanique [...] j'ai l'honneur de vous informer que Son Excellence le Ministre de la Marine, par une détermination du 1er septembre, a bien voulu fixer ma résidence à Paris en activité de service. Si, dans la suite, et pour des circonstances impossibles à prévoir, on m'offrait des fonctions à remplir incompatibles avec ma résidence habituelle à Paris, quels que fussent les avantages de ces fonctions, j'y renoncerais sans hésitation, plutôt qu'à l'honneur de siéger à l'Académie des sciences. C'est ce dont je prends l'engagement formel.
2. Communiqué from Dupin (1820) to the Minister for the Navy (ACSM):
   Si l'industrie des peoples voisins restait stationnaire, nous pourrions abandonner la nôtre à la lenteur de ses progrès naturels et peu nous inquiéterait de voir la routine arrêter la marche de ces progrès. Mais à l'ouest, au nord, à l'est de la France, des nations riches et populeuses s'occupent sans cesse d'accroître les moyens de perfectionner leurs arts utiles.
3. From Dupin's *Compte rendu, Forces électorales à la fin de 1827* (Dupin 1827):
   En 1827, la France a les plus grands efforts à faire pour s'élever par l'enseignement élémentaire au simple niveau des peuples que nous regardons comme ignores. A cet égard, nous sommes au-dessous des Irlandais et des Autrichiens; sept millions de Français savaient lire il y a quarante ans; douze millions le savent aujourd'hui; vingt-six millions devront le savoir. Travaillons pour offrir à quatorze millions de nos concitoyens la première et la plus indispensable des

connaissances. Il faut, en dix ans, faire plus que nos pères n'ont fait en quarante.

4. From Dupin's letter to Pasteur, 28 August 1872 (ASHM):
Mon cher et illustre confrère, permettez-moi de vous adresser mes félicitations les plus vives au sujet de vos admirables recherches sur la fermentation alcoholique.
Tout se réunit pour faire de ce travail l'époque d'une ère nouvelle de la science à laquelle vous devez votre juste renommée. J'ai l'honneur de vous saluer avec admiration.
5. Letter from the Comte du Hamel de Breuil, announcing the death of his father-in-law, Dupin, 18 January 1873 (AAS):
Mon beau-père a succombé ce matin […] des suites d'un accident survenu par la rupture d'une sonde dans la vessie […] M.Dupin est mort en pleine connaissance, patient, résigné, réconforté par la foi et la confiance dans la miséricorde de Dieu.
6. Anonymous verse (n.d) on the subject of Dupin *aîné* (ADN 4 J, packet 1/2):
Tout pouvoir, tour à tour, peut dire: il est des nôtres,
Du pain dur aux proscrits, du pain mollet aux autres,
On lui présente un siège; il n'est pas indécis,
Et tout Paris se dit: c'est du pain rassis.
De la fidélité, ce n'est qu'un faux semblant,
Il fut bleu, rouge même … il serait du pain blanc,
Ce digne magistrat, offrez-lui quelque lucre,
Et d'aigre qu'il était, voilà du pain de sucre!
7. Letter of 27 February 1814 from Charles Dupin in Toulon, to his elder brother (DRB):
Il paraît que M. notre père a des raisons particulières et spéciales de ne pas m'écrire. Pas de nouvelles depuis un an.
Reply to his brother, who claimed that their father was very busy:
Je ne prends pas des vessies pour des lanternes (Ibid.).
8. Letter of 21 October 1841, two years before his death, from Charles Dupin's father to his son (DRB):

Si tu considérais mon grand âge, et l'accident funeste qui est venu aggraver et précipiter le cours de ma vie, nous profiterions au moins du peu qui me reste pour rendre nos relations plus fréquentes et d'autant plus consolantes. Tôt ou tard, tu éprouveras à ton tour tout ce que redoublement d'affections filiales procure de satifactions aux pères [...] Souviens-toi que ce qu'il y a de plus difficile dans la vie, c'est de la finir et de la bien finir aux yeux de Dieu et des hommes.

9. Document concerning the marital situation of the Dupin parents (AMV):
   Le père des trois Dupin, originaire de Clamecy, y faisait sa résidence habituelle [...] mais par l'effet d'une dissidence intérieure qui, à un moment donné, finit même par une instance de divorce [...] la mère vivait dans une séparation de fait d'avec son mari. C'est par suite de cet état d'éloignement de fait qu'elle était successivement accouchée à Varzy de ses trois fils.
10. Dupin père to his eldest son, 30 August 1804 (ADN), about his wife's behaviour:
   C'est bien sot. C'est être aussi bête de méchant. C'est vouloir être absurde, même à ses dépens. J'ai tout fait pour n'avoir rien à me reprocher. [...] M. le grand ingénieur ne vaut pas mieux.
11. Ibid., 15 July 1818 (DRB):
   J'ai commencé à lire les mémoires de Charles. Il est fort quand il est sur son terrain. Sa réputation est là, plus que dans ses excursions en littérature. Ayant trop longtems dédaigné tout ce qui n'était pas des chiffres et des lignes, il écrit dans sa langue maternelle comme un étranger qui vient de l'apprendre [...].
12. Extract from the sixth of nine letters from Dupin to his friend Mathieu (ADN Dupin 4J). This sixth letter is dated from Genoa, 10 May 1806):
   J'ai proposé à Hachette de faire pour l'Ecole polytechnique un ouvrage dont voici le plan [...] Entre les connaissances d'analyse appliquée des aspirans à l'école polytechnique et le cours deMonge, il existe un espace immense, dont le cours de Hachette ne remplit

que la plus petite partie: Hachette ne traite que légèrement les seules surfaces de second degré. Il n'en considère que les plans tangents et fait abstraction de leur courbure, et son cours devrait faire le germe de toutes les grandes théories du cours de Monge; en arrivant à celui-ci, et avant d'appliquer aux surfaces le calcul des différences partielles, ils devraient savoir tout ce que le calcul différentiel simple apprend sur les surfaces et Hachette ne dit rien de tout cela. Aussi vous savez combien peu d'élèves sont en état de suivre le cours de Monge.

13. Letter from Dupin to Mathieu, 10 May 1806 (Ibid.):
Je voudrais savoir si ce coquin de Monge sera assez brave homme pour me répondre sur mon travail "Des déblais et des remblais"; je jurerais que non. [...] Songez que pour trois sous s'affranchissent les lettres aux militaires et que la vôtre coûte ici 18. Songez que les militaires sont des gueux sans monnaie. Affranchissez-vous ou envoyez vos lettres à mon père qui me les fera parvenir sans payer de port.

14. Letter from Dupin to Mathieu 13 *brumaire an* XII (23 December 1803) (*Ibid.*):
Ce sublime Bailly écrit avec tant d'intérêt et de charme; c'est avec tant de noblesse et de génie qu'il expose les grands phénomènes de la nature et l'histoire de leurs découvertes qu'on ne peut se lasser de l'admirer, de le lire et de le relire. Vous en tirez tout le charme du style, toute la beauté des idées et le mérite de ses ouvrages quand vous les comparez avec ceux des autres hommes [...] Le "système du monde" de La Place, par exemple, vous plaira presque toujours, et, quelquefois même, vous élèvera jusqu'à l'enthousiasme si vous le lisez sans connaître Bailly. Mais si vous le connaissiez déjà, le charme tomberait à l'instant. Vous aimeriez la dignité plus simple, plus naïve et cependant plus éloquente et plus majestueuse de Bailly.

15. The following letters to Dupin from Mathieu are in ADN (fonds Dupin, 4J):
Gênes, 22 février 1806
Ce diable de Monge est absolument introuvable et homme à ne me donner aucune réponse. Eh bien! que voulez-vous mon cher Mathieu,

nous y renoncerons et puis cela sera fini. Je n'aurai qu'un regret en tout cela, c'est de vous avoir donné autant de peine inutilement. Puisque vous êtes dans les Ponts-et-Chaussées, vous devez avoir le Mémoire de Monge sur les déblais et remblais, Académie des sciences, Paris, 1781. J'ai beaucoup travaillé sur ce sujet depuis mon séjour à Gênes. Si cela pouvait vous faire plaisir, je vous donnerais pour vous mon travail et vous pourriez, s'il vous en paraissait mériter la peine, le montrer à Prony que ce sujet doit intéresser par état. Mais, comment diable pourrai-je jamais savoir cela? Vous êtes un homme qui n'écrivez pas et, dont on ne peut avoir de nouvelles que par voie tierce. Convenez au moins que ce n'est pas là votre bon côté.

16. Toulon, 17 mars 1806:
    Vous m'aviez dit autrefois avoir communiqué l'analyse de mon mémoire sur les 'déblais et remblais' à M. Delambre. Probablement, il l'aura perdu de vue et ne vous en aura jamais parlé.
    A propos de cette analyse, j'ai été tout étonné, il y a deux à trois jours, lorsqu'un de mes collègues, lisant 'le Moniteur' me cita un article ainsi conçu: Corresp. Polyt; (sic). Analyse d'un mémoire de M. Dupin sur les déblais et les remblais [...] etc.
    J'avais donné cette analyse à Hachette afin de la donner à Monge, quand Hachette et Poisson vinrent à Gênes. Parce qu'Hachette, à qui j'avais envoyé, pour Monge, une première copie, m'avoua ingénument qu'elle était perdue et qu'il ne savait pas ce qu'il en avait fait: ce sera probablement sur la deuxième copie qu'Hachette aura fait son article qui me concerne'.
17. Three more letters from Dupin to Mathieu written in 1806, and from which I provide extracts, are again dated from Genoa (*Ibid*):
    Gênes, 27 mars 1806
    Vous remarquerez peut-être, mon cher Mathieu, que j'ai ramené l'optique à n'être plus qu'un cas particulier de la théorie des Déblais et des remblais. Vous avez dû voir dans le n° 5 de la Correspondance une belle propriété d'optique donnée par M. Malus. Il y est parvenu par l'analyse; je l'ai démontrée par la géométrie; je l'ai poussée tant

que je l'ai pu. […]. Je suis revenu beaucoup sur ce que j'avais fait sur la courbure des surfaces. Imaginez-vous, mon ami, que toute l'optique, toute la coupe des pierres, toute la théorie des déblais et des remblais, etc., ne forment, avec la théorie des signes de courbure, qu'un seul et vaste système d'idées […].
Il faudrait pendre tous ces coquins de sénateurs invisibles et affairés, et ce sont pourtant les meilleurs hommes de la Terre; Je n'ai pas besoin de dire que je suis flatté de l'opinion de M. Monge. J'aurais cru, pourtant, que l'idée de chercher trois systèmes de surfaces constamment rectangulaires entre elles était celle qu'il aurait préférée. Voici ce que je crois avoir démontré:
Si trois systèmes de surfaces quelconque sont tels qu'elles se coupent trois à trois à l'angle droit à chaque point de leur intersection, comme seraient un ensemble de petites cases cubiques, les surfaces du premier système sont coupées par celles du second suivant toutes leurs lignes d'une des courbures et les mêmes surfaces du premier système sont coupées par celles du troisième suivant toutes les lignes de l'autre courbure; ainsi, dès qu'on connaîtra un pareil ensemble de systèmes, qui comprendra une surface donnée, les intersections de cette surface et de celles des autres systèmes donneront les lignes de courbure.

18. A letter of 10 May from Genoa, Dupin to Mathieu (ADN):
Voici un exemple de la simplicité que je m'efforcerais de mettre dans les sujets les moins faciles: Supposez qu'on me demande de déterminer la direction des lignes de courbure d'une surface du second degré et les rayons de courbure de la même surface pour un de ses points donné. Voici comment j'y parviens:
Par le centre de la surface, je mène un plan parallèle au plan tangent à la surface au point donné. Je détermine les axes de la section de ce plan dans la surface. Le grand axe est parallèle à la tangente de la grande ligne de courbure au point donné; le petit axe est parallèle à la tangente de la petite ligne de courbure au point donné; le quarré du petit axe de la section, divisé par la distance du centre au plan tangent au point donné, égale le plus petit rayon de courbure.

Et ces résultats s'obtiennent par des considérations immédiates; ils ramènent, comme vous le voyez, les contacts du second ordre à des déterminations proportionnelles, ce qui est bien plus facile.
J'ai pioché comme un malheureux la courbure des surfaces. Tout ce que j'ai trouvé sur les surfaces du 2ème degré, j'ai cherché à le généraliser pour des surfaces quelconques et cela m'a conduit à quelques résultats que je vous ferai connaître par la suite. Je puis résoudre immédiatement et graphiquement, par exemple cette question-ci: Trois courbes quelconques se croisent au même point sur une surface; on connaît en ce point la courbure de ces courbes; on demande la direction des lignes de courbure de la surface et la grandeur des deux rayons de courbure qui appartiennent à ce point de la surface. Cette solution peut servir dans la coupe des pierres, qui, comme vous le savez, est intimement liée à la théorie des lignes de courbure.
Quand je vous ai parlé des Mémoires de l'Institut au sujet de mon travail sur les Déblais et les remblais, c'était bien plus pour savoir l'opinion de M. Delambre que dans le désir de voir mes mémoires dans les savants étrangers, honneur assez médiocre en lui-même.

19. From Dupin to Dupin *aîné*, from Toulon, 8 December 1806 (DRB):
Je suis arrivé après 23 jours de route et avoir fait 24 lieues à pied. J'ai été vingt jours à bord, couché sans matelas sur la planche, mais je me porte bien; cela est passé, je n'y pense plus.
J'ai mille compliments à te faire, mon cher ami, pour ton avancement. Loin de me surprendre, il remplit toute mon attente et je remercie seulement la fortune d'avoir été juste pour toi.

20. From Dupin to M. Baron, commandant supérieur de la Marine (ADN):
Mon commandant,
D'après vos ordres, j'ai pris connaissance précise de nos ressources en toile à voile, afin de voir si nous pourrions faire face à la demande de 14 000 [illegible] faite par le Génie de terre. Il est de toute impossibilité d'y faire face. En effet, nos toiles à voile font partie des voiles neuves restant des Russes et que le bien du service ne peut permettre de sacrifier entièrement. Le reste tombe en lambeaux et est hors d'état de remplir aucun service.

21. From Dupin to his elder brother from Toulon, 12 November 1813 (DRB):
Un petit billet pour M.Greffier imprimeur [...] Je tiens beaucoup à avoir 100 exemplaires de l'analyse de mon travail sur le bois, puisque ce travail paraît assez bien reçu des savants et du Ministère. Il sera plus lu des ingénieurs que des théories moins faciles et plus abstraites. J'ai envoyé dernièrement le même travail à Poisson et il m'a répondu qu'il l'avait aussitôt envoyé à l'imprimerie impériale pour être imprimé dans la collection du "Journal de l'Ecole polytechnique" où les premiers savants se font l'honneur de mettre leurs productions.
22. Dupin to Mathieu, 7 April 1810 (ADN):
Je vais te donner une petite idée du volume académique que j'achève de composer et que je vais
envoyer à M. Lacépède pour en faire hommage à l'Institut.
1) Histoire de l'Académie ionienne pendant la première année, par Dupin.
1bis) Rapport de M. Massapoulo sur les opuscules de M. Dupin sur l'enseignement et l'étude du droit.
2) Comparaison des poids et mesures françaises avec celles des sept isles et de l'Albanie, par Augoyat, Dupin et Arnaud, ex-élèves de l'Ecole polytechnique.
3) Mémoire sur l'établissement de la vaccine dans les sept isles par le D Gangadi des Académies italiennes et des Georgophiles de Florence.
4) Mémoire sur le traitement des fièvres quartes, par le D. Franck de la Commission des sciences et des arts d'Egypte.
5) Mémoire d'Augoyat sur les pierres meulières artificielles employées dans la Grèce et les isles
ioniennes.
6) Mémoire sur les beaux-arts par M. Deluret, ingénieur militaire.
7) Eloge du Grec moderne Boulzaris qui a traduit tout Virgile en vers grecs homériques, par le

D. Marromatis. Eloge écrit en grec ancien, traduit en français par Dupin.

8) Hymne à l'Empereur Napoléon en vers grecs pindariques, par le même.
9) Ode italienne sur le passage du Danube et la bataille de Wagram, par M. Delviniotti de
l'Académie italienne.
10) Traduction du premier chant du poète grec Quintus Smirneus, continuateur d'Homère, par le
même Delviniotti.
11) Traduction en grec vulgaire de la conspiration de Catalina de Salluste, par M. Capo d'Istria.
12) Note sur deux belles collections de médailles grecques et romanes existantes [sic] à Corfou
et contenant plusieurs médailles précieuses et d'autres nouvelles, par M. Pro Salendi, antiquaire
connu de [illegible word] et qui se propose de faire un grand travail sur ces deux collections.

23. Note from Dupin to Forestier, chef de la 1ère division de Ministère de la Marine et des Colonies, dated from Pisa, 5 October 1812 (ASHM).
Vous connaissez mes malheurs; partis quatre de Corfou, mes trois compagnons de voyage sont morts dans un espace de quinze jours; et après quinze mois de souffrances, je ne jouis encore que d'une santé languissante [...].
24. Letter from Dupin to his father, dated from Naples, 27 December 1811 (DRB).
Enfin, le charme s'est levé et, pour la première fois depuis tant de mois, je puis recevoir une lettre de vous [...]. La kinkina m'a fait éprouver les suites que vous craigniez pour moi; j'ai des obstructions, mais elles ne me font pas souffrir et elles cèdent un peu aux remèdes avec lesquelles je les combats.
Comment! si je sais assez d'italien pour me faire entendre? Vous ne scavez [sic] donc pas que j'ai vécu cinq ans au milieu des Italiens et

fort souvent au milieu de sociétés purement italiennes, et que nous parlions et écrivions toujours en italien dans l'Académie ionienne? et que je suis une espèce de petite puissance italico-littéraire? Peste! un homme qui compte sept à huit sonnets italiens composés en son honneur! Ici, je n'y tiens plus et il faut que je vous découvre toute ma gloire.

La fortune, lasse enfin de me persécuter, m'offre en arrivant à Naples, un médecin qui ne guérit pas ses malades, mais qui les chante […]. A-t-il expédié quelque épouse, il en fait l'oraison funèbre en vers élégaiques. Il redit comment les potions, les calmants, les excitants et la rhubarbe et le sené, rien n'a pu l'empêcher de boire aux eaux fatales du Léthé! Il redit avec quelle patience elle prenait les médicaments les plus amers et ses vertus et sa confiance en l'infortuné médecin.

Alors, il pleure sur ses vers avec le pauvre époux et ils se consolent ainsi peu à peu dans les douceurs d'Hippocrate et des Muses.

25. From Sané to Vice-Admiral Leysseguer, 19 February 1813 (ASHM).

    La demande […] ne saurait être exaucée, car Dupin vient d'être nommé au commandement d'une compagnie d'ouvriers militaires à Toulon. […] les ouvriers miltaires en poste à Corfou, en raison de leur mauvaise discipline, devraient être remplacés.

26. From Dupin *père* to his eldest son Dupin *aîné*, 26 January 1812 (Extract, DBR).

    J'ai reçu une lettre de Charles et n'en suis pas content. Ses lettres, fort rares, fort courtes, ne parlent que de littérature ou de balivernes; trop peu de choses sur sa santé et sur son régime, rien du tout qui soit relatif à sa profession d'ingénieur. On dirait qu'il a donné sa démission ou qu'il ne craint pas d'être considéré comme démissionnaire.

27. From Charles Dupin to his father, dated from Naples 10 March 1812 (extract, DRB).

    Je sollicite la place de secrétaire des constructions maritimes, séant à Paris, place qui ne peut être remplie que par un sous-ingénieur et ne m'ôte pas de mon poste.

Cette place est parfaitement conforme à mon goût pour l'étude. Il y a plus: je sens tellement pour moi le besoin d'habiter Paris, afin d'y recueillir le fruit de mes travaux et de mes connaissances, que si, dans un an, je n'obtiens pas quelque fonction qui m'y fixe, je donnerai ma démission pure et simple et j'irai vivre à Paris sur mes petites épargnes en attendant que je trouve une place.

Mais combien ne serait-il pas plus avantageux de rester dans mon Corps, d'avoir un poste où l'on peut développer quelques talents et de vivre auprès de mes frères et de ma famille.

Recommandez-moi donc à vos anciens amis afin qu'ils écrivent en ma faveur, soit au ministre ou à M. de Sané, ou à M. le vice-amiral Rosily, président du Conseil de constructions.

Adieu, mon cher papa, répondez-moi à Rome, où je compte aller sous quinze jours, ma santé allant un peu mieux [...].

28. From Dupin père to his eldest son, 2 April 1812 (DRB):

Je t'envoye, mon ami, une lettre de Charles. Il faut tout de suite en conférer avec M. Carnot et M. de Sané;

Voir 1°) le quid utilius

2°) la possibilité et les moyens de succès; agir, s'il y a lieu, sinon, donner de bonnes raisons à notre jeune homme; en un mot, faire au moins preuve de zèle et d'attachement. Il te sera facile d'arriver à M. de Rosily par une de mes anciennes connaissances.

29. From Dupin père to his eldest son, 15 June 1812 (DRB):

Charles m'écrit, sur la tour de Pise: j'aimerais mieux le voir sous celle de Saint-Martin. Il va mieux et me fait espérer qu'il démarera après une nouvelle quarantaine. Dieu veuille qe ce soit la dernière.

30. From Dupin *père* to Dupin *aîné*, 3 December 1812 (extract, DRB):

Je regrette beaucoup que ton frère ne soit pas logé chez toi. Veille du moins à ce qu'il ait chez lui le nécessaire et surtout du feu car, dans un convalescent, le refroidissement peut amener des rechutes fâcheuses. Occupe-toi de lui pour ce qu'il ne s'occupe pas assez, c'est à dire de lui-même et de ses vrais intérêts.

31. From Dupin *père* to Dupin *aîné*, 4 January 1813 (extract DRB):

Charles ne m'écrit jamais rien de ce qui le concerne individuellement, ni par rapport à son état. Que disent ses médecins? Je ne trouve rien de plus inconvenable qu'un scavant [sic] qui ne fait rien savoir de ce qu'on voudrait savoir à moins qu'on ne l'interroge en personne.

32. Dupin *père* to Dupin *aîné*, 20 March 1813 (DRB):
Je m'inquiète bien que ton mathématicomane devienne membre correspondant de l'Institut tant qu'il ne sera point, avant tout et par dessus tout, membre correspondant de sa famille. Je viens pourtant d'en recevoir enfin une lettre qui en suppose une autre apparemment restée dans son fouillis de papiers et, au surplus, remplie de ses lubies ordinaires qui ne sont pas ce que je voudrais trouver dans les lettres de mes enfans [...]. Il rend ses mathématicailles insupportables à force d'en parler.

33. From Dupin père to Dupin aîné, 9 July 1813 (DRB):
J'ai reçu la gravure de Charles. Jamais je n'ai vu un portrait moins ressemblant [...] Tous tant que vous êtes, vous n'entendez rien à vous faire peindre et vous croyez vous reconnaître dans tous les portraits pourvu qu'il y ait un nez et des oreilles [...] Je crains fort que Charles ne quitte son état ou n'en soit renvoyé pour sa mauvaise tête et que Philippe ne soit interrompu dans le sien. Ainsi, tous les deux sont la cause occasionnelle de mes maux d'estomac qui vont toujours en croissant.

34. From Charles Dupin to his elder brother, 26 October 1813 from Toulon (DRB):
Mon cher Dupin,
[...] quant à Philippe, je l'excuse moins et je ne m'en plains pas, car il est jeune encore; il apprendra le prix de l'amitié lorsqu'il aura besoin des hommes.
J'ai parlé froidement du domaine de Cœurs à mon père; je serai toujours également froid sur ce qui regarde les intérêts matériels. Mon cher Dupin, de l'indulgence et point d'humeur entre nous; tu ne m'aimeras jamais autant que je t'aime.
Ton frère, Charles Dupin.

35. From Charles to Dupin *aîné* (extracts) (DRB):
J'aurais beaucoup mieux aimé, je l'avoue, que tu eusses vu toi-même le ministre de la Marine. Je n'ai jamais pu savoir si tu avais reçu la lettre de présentation que j'avais pris la précaution de t'adresser pour le ministre par le moyen du Conseil des constructions. As-tu reçu cette lettre? L'as-tu jointe à mon ouvrage? J'ai déjà fait l'analyse de l'ouvrage de Monge; je la joins à ma lettre et je te prie de la porter à ce digne homme. C'est bien le moins que tu puisses faire pour le remercier de son amitié et ses attentions pour moi et pour toi. [...].
Donne-moi l'adresse d'une personne à Auxerre à qui je pourrai adresser les oranges de mon jardin qui commencent à mûrir et quelques autres productions de la Provence. [...] Ce que tu me dis relativement au refus de payer de MM. les Parisiens m'embarrasse fort, parce que j'ai 2 000 F à t'envoyer et je ne sais comment faire pour que cela soit sûr. [...] Tu serais bien aimable si tu voyais un peu Madame Courcier. Il y a 20 pages sur Démosthène tirées depuis six mois.
P.S. Si je vais à la Grande Armée, je ferai l'impossible pour passer par Paris et te serrer encore sur mon cœur.
Si tu trouves ma lettre trop chère, tu la porteras sur mon compte, pour ne pas se gêner, il faut agir librement entre frères.
36. Ibid. 23 November from Toulon (DRB):
J'ai reçu de Mme Courcier la fin de ma dernière harangue de Démosthène et le commencement de mes "Considérations". J'ai fait beaucoup de changements en profitant de tes bons conseils.
37. From Dupin to Bajot, *chef du Bureau des Lois au Ministère de la Marine*, from Dunkirk, 18 March 1816 (ADN fonds Dupin 4J):
Je vous adresse une description des sculptures de Bajot qui a été très bien accueillie par l'Institut. Si elle vous convient, vous l'insérerez dans tel de vos numéros que vous voudrez. Je vous donnerai ensuite la description du musée maritime que j'ai établie dans l'arsenal de Toulon. Voudrez-vous, S.V.P., demander à M. Portier si, comme il me l'a promis, il a fait son rapport <u>sur le voyage que je demande à faire en Angleterre.</u>

38. From Dupin to Mathieu, Dunkirk 3 April 1816:
Mon ignorance est complète sur les nominations aux places d'académicien libre. Faut-il être vicomte, marquis, vidame ou géomètre, physicien, astronome, je n'en sais rien.
39. Dupin to his elder brother, Dunkirk, 5 April 1816 (DRB):
Huit places d'académiciens libres sont créées à l'Académie des sciences par la nouvelle organisation. Je sollicite d'être mis au nombre des candidats pour l'une d'elles. J'ai écrit pour cela à M. Delambre, secrétaire. Si tu pouvais voir MM. Dagueux et Lefèbre-Gineau pour prendre des renseignements sur le mode et l'époque certaine de l'élection, afin de me l'apprendre sur le champ, tu me rendrais grand service. J'obtiendrai une permission de 15 jours pour aller à Paris faire valoir mes titres (la permission me sera donnée ici par le général Goux et me mettra parfaitement en règle). Ma nomination à l'Académie changerait la face de mes affaires et me ferait certainement placer à Paris par le Ministre de la marine.
40. Dupin's "Plan de voyage en Angleterre", January 1816 (ASHM CC7 779) :
Aller en Angleterre observer l'état des Arts directement ou indirectement relatifs à la Marine et décrire:
    1) Les agents et les moteurs employés par les Anglais pour exécuter les travaux de ces arts.
    2) Les instruments et les outils employés pour donner aux objets les formes et les qualités qui leur conviennent.
    3) La suite des procédés méchaniques [sic] et géométriques employés pour faire passer la matière première de l'état brut à l'état de produit industriel.

Considérer ensuite l'application de ces moyens à la construction, au gréement, à l'armement des vaisseaux.

Jamais étude de ce genre ne fut plus intéressante pour la France que celle de la Marine anglaise; jamais moment ne fut plus opportun. Une guerre longue et fertile en grandes expéditions; toutes différentes en but, en moyens, en résultats, a créé dans les arts maritimes des besoins inconnus jusqu'alors et des ressources inespérées pour satisfaire à ces besoins.[...] Ainsi, la léthargie de la paix, qui rend l'art stationnaire dans nos arsenaux oisifs nous offrira les moyens de perfectionnement les plus avantageux, pour une comparaison impartiale et raisonnée de nos travaux avec les travaux de nos émulés.[...] Il existe en France deux opinions également erronées, également pernicieuses. Ceux qui partagent la première croient que nos travaux nautiques ont atteint le maximum de perfectionnement ... Ceux qui partagent la seconde opinion croient au contraire que tout est parfait en Angleterre.[...] Je ne crois pas à notre perfection en marine; je crois encore moins à celle des Anglais

Disons un mot de nos moyens d'exécution. C'est comme géomètre et comme

physicien que je parcourrai l'Angleterre. C'est avec la recommandation de l'Institut de France et de plusieurs académies et des ingénieurs anglais. Nos amis et ceux de ma famille me fourniront encore d'autres facilités et, malgré l'esprit mystérieux qui s'opposera plus d'une fois à mes recherches, je suis sûr d'en voir assez pour acquérir une juste idée des établissments où je ne pourrai pas pénétrer.

41. Dupin *père* to Dupin *aîné*, 14 July 1816 (DRB).

Donne à Charles toutes les instructions pour se bien porter et se bien comporter en Angleterre, pour ne pas s'exposer à l'Alien Bill ni à Bedlam. Bogne prétend qu'il ne verra rien parce que l'on ne montre rien aux étrangers.

Recommande-lui de ne point abuser de la liberté de parler, pour ne point se compromettre ni avec la police anglaise ni avec la police française.

Je désire qu'il s'en revienne avec la tête froide comme les glaciers de la Suisse et avec une tempérance de conversation comme Jacques Rosbif. Souhaite-lui bon voyage de ma part.

42. Dupin to Bajot, 21 July 1816 (DRB):

D'ici à un mois, je vous ferai passer copie d'un rapport que l'Institut vient de faire sur un de mes mémoires de physique mathématique [...]. J'ai reçu mes papiers hier et je pars demain [...]. On m'a donné 1 200 F d'indemnité pour six mois et j'emporte 3 000 pour trois mois.

43. Letter from Charles to his elder brother, from London 2 August 1816 (ASHM):
    Londres, 2 août 1816
    Je suis arrivé à Londres en bonne santé. J'ai pris une chambre pas trop chère et je vis avec ma donation, moyennant quoi je me tirerai d'affaire.
    Je vois de toutes parts des choses admirables dans une ville de quatre lieues de long sur dix de large. Les habitants qu'on nous représentait comme des hommes qui nous haïssent, sont d'une grande complaisance et ne nous injurient jamais.
    Voici mon adresse: Mr Dupin, at M. Edden, taylor and draper, Christchurch.
44. Dupin to his friend Mme Prony, 2 August 1816 (BN Naf 3282):
    Comme toutes les joies de ce monde sont courtes et transitoires, je vais quitter la plus jolie habitation du monde, où j'avais à ma droite les prisons de Newgate, en face une église, à côté un cimetière planté d'arbres en quinconce et parsemé de pierres tombales! enfin à gauche un hôpital. Voilà, Madame, une demeure digne du sombre Young et du lugubre Hervey, et si jamais Mademoiselle Lucie vient visiter l'isle de la mélancolie, elle ne pourra trouver nulle part un réduit plus sentimental. Quant à moi que rien n'attriste, pas même les éternels nuages de charbon de terre, qui sont pour moi ce qu'Alexandre était pour Diogène, mais qui n'ôtent jamais de devant moi mon soleil, la solennité des objets qui m'environnent ne m'impose pas le moins du monde, et n'ajoute rien à ma gravité.
    Après avoir admiré Mesdames les Anglaises qui ont toutes un joli visage, un beau sein, de gros pieds, un gros dos et le col tendu, je vais admirer les provinces. J'avais envie d'acheter des culottes et des guètres vertes, avec un habit descendant jusqu'aux talons pour m'empêcher d'avoir un air étrange, mais réfléchissant que ma manière de parler, et surtout de comprendre ce qu'on me dit, c'est à dire, de n'y rien entendre, détruiraient tous les prestiges de ma toilette, j'ai renoncé à tant de prétensions.

Il faut que je compte beaucoup sur votre indulgence pour ne pas craindre que mon bavardage ne vous excède, tandis que je pourrais vous parler de choses plus intéressantes de m'arrêter sur de semblables niaiseries.

Je vous prie de me rappeler au bon souvenir de l'excellente Madame Girard; c'est une ancienne amitié, et il n'y a qu'en amitié où je crois à la douceur et à la plennité des souvenirs. Dussent Madame de Stael, Monsieur de Chateaubriant et les plus aimables demoiselles du monde prononcer anathème sur cette manière triviale et anti-romantique de peser les manières [illegible] affections de la portion sentimentale de notre être.

Voilà pourquoi j'ai probablement peu de dispositions pour écrire des romans quoique ma vie en soit un, ou plutôt en referme une demi-douzaine. Veuillez, s'il vous plaît présenter mes civilités à Madame Gail, qui me semble avoir presqu'autant que moi des dispositions au recueillement et à la morosité et dire à Mademoiselle Lucie, que malgré mes méchancetés, personne plus que moi n'admire sa mérite et ses qualités : elle est digne d'être votre élève.

J'ai l'honneur, Madame, de vous saluer avec le plus entier devouement.

45. From Dupin in London to the Minister for the Navy, 1 Septembre 1816 (ASHM):

    Monsieur l'Ambassadeur de France, auquel vous m'aviez recommandé, a pris le vif intérêt au succès de mes sollicitations, et il est allé lui-même deux fois à l'Amirauté d'Angleterre; il a écrit en ma faveur.

    A présent, Monseigneur, vous pouvez vous en rapporter à mon amour pour les Arts et les Sciences, du soin que je vais mettre à rendre mon voyage fructueux.

46. 1 September 1816, Dupin to an Unspecified person (ADN):

    J'ai réussi dans tout ce que je voulais ; j'ai l'autorisation [...] de voir tout ce que je désire observer.

    Je crois qu'aucun voyageur n'avait encore été traité d'une manière aussi favorable.

47. On 2 October, 1816, Dupin provides more details and complains to his elder brother about his lack of funds, as he often did (ADN):
Portsmouth, le 2 octobre 1816
Je suppose que tu vas lire mon itinéraire détaillé dans ma lettre à mon père. Tu verras que je ne fais pas le paresseux et que, si je reviens sans être bien instruit, ce ne sera pas faute d'avoir bien étudié.
Force m'est de voyager économiquement pour ne pas manger en trois mois ce qui doit m'amener jusqu'à six.
Aussi me suis-je emparé de l'impériale des diligences anglaises … j'y ai pris place dans un coin honorable et avantageux, à droite et sur le devant pour voir le pays et avoir ma main droite libre. Je voyage avec habit, redingote, capote et parapluie, mon dictionnaire dans une poche et la description alphabétique des villes et des comtés dans l'autre poche.
48. Dupin to his elder brother, London, 5 November 1816 (ADN):
Voilà près de trois mois que je n'ai pas de tes nouvelles et tu conçois combien j'aurai de plaisir lorsque j'en recevrai. Nous avons six courriers en arrière par suite du mauvais temps. Peut-être m'apporteront-ils quelque chose de vous.
J'ai encore besoin de toucher 800 des 1 200 que tu as remis à M. Lafitte. Tu devrais bien dire à mon père que, si jamais il doit se déterminer à me faire le fameux cadeau dont il parle depuis si longtemps, c'est à présent que 5 ou 600 F me mettraient par conséquent à mon aise.
49. Dupin to his elder brother, 23 January 1817 (DRB):
J'ai quitté Dunkerque le <u>22 juillet 1816</u> et j'y suis rentré le <u>23 janvier 1817</u>, ce qui me fait six mois juste. Je te prie de régler l'état de mes dépenses. Je vais toucher ici une petite somme qui m'est due et avec laquelle je te paierai sur le champ, en te remerciant, mon bon ami, des avances que tu m'as faites. [….] Je reviens, mon cher ami, avec une riche collection d'observations, de plans, d'expériences et <u>trois mois me suffiront à peine pour les mettre en ordre et les rendre présentables</u>. […] l'ambassadeur de France, auquel j'avais été recommandé très froidement et qui m'avait reçu en conséquence, a fini par prendre intérêt à mon voyage lorsqu'il en a vu le succès.

Il m'a ouvert sa maison et m'a souvent invité à dîner chez lui. A mon avant-dernière visite, il me demanda si l'avocat du Maréchal Ney m'était parent.

Frère, M. l'ambassadeur.

Il a un très grand talent.

Vous êtes très bon.

Mais je crains qu'il n'ait des principes trop républicains.

Au contraire, Monsieur, et je l'ai souvent vu manifester avec véhémence contre ce système; il était d'ailleurs fort mal vu du Gouvernement provisoire de Buonaparte.

Eh bien! Vous me faites plaisir et je suis charmé par ce que vous me dites. [....]

Un moment après, une Lady Clainger, avec laquelle j'avais dîné chez l'ambassadeur, me demanda tout haut si j'étais son frère.

Oui, Madame.

Eh bien! Monsieur, je suis enthousiaste de votre frère; j'étais présente à la défense des trois anglais; elle lui fait un honneur infini et je l'aurais embrassé si je l'avais pu. Dites lui cela de ma part.

Je t'avais écrit pour te prier d'envoyer, par la diligence, 40 exemplaires des feuilles de
mathématiques dont M. Gergonne, de Nîmes ou de Montpellier, m'a écrit t'avoir adressé 100
exemplaires. (Document Romain Baron).

50. Dupin to Prony (n.d. ENPC, ms 1877):
Sans votre recommandation à M. Rennie et celle de M. Rennie à M. Whidbey, je n'aurais jamais pu donner une idée aussi étendue et j'ose dire satisfaisante d'un ouvrage qui fait honneur à M. Rennie [...] je poursuis mon voyage, plein d'admiration pour une foule de belles choses qui frappent partout mes regards.

51. Extract from an application for a patent by John Walters, 1815 (Office de la propriété industrielle,
CNAM):
Description des moyens d'execution et dessins relatifs à la demande de John Walters Architecte et Ingénieur de Fenchurch Buildngs dans

la Cité de Londres, tendante à obtenir un brevet d'importation, pour un nouveau genre de perfectionnement dans la construction, aussi bien que dans le radoub des vaisseaux, perfectionnement dont il est à la fois l'inventeur en Angleterre et l'importateur en France.

Dans une coupe perpendiculaire à la longueur du vaisseau qui doit être construit ou radoubé, et quelque part auprès du centre de gravité, j'introduis entre les coupes de levée, un couple principal A (Fig 1 et 2 ou 3) attaché à la quille, ayant à l'intérieur des bras diagonaux B (Fig 2 ou 3) et étant attaché au travers du vaisseau par une ou plusieurs barres ou baux de lien (C. Fig 2 ou 3) à la hauteur du pont ou des ponts.

52. 9 February 1817, Dupin to Dupin *aîné* (DRB):

J'ai ici deux mandats de 660 F. Je garderai 60 F qui, avec les 50 autres que j'ai dans ma poche, me fourniront un capital pour boire et manger jusqu'à nouvel ordre. Je vais songer aux moyens de t'envoyer tes 600 F. Tu les recevras incessamment [...]Tu verras que, quand tu auras reçu les 600 F que je vais t'envoyer, mes dettes seront plus que payées. Le reste sera pour solder une petite somme que je dois à Mme Courcier.

53. 9 February 1817, Dupin to Dupin *aîné* (DRB):

J'ai reçu une lettre fort agréable de l'amiral Rosily, à qui j'avais adressé la description des travaux entrepris pour lever les grandes cartes de l'Angleterre. Comme l'amiral Rosily est à la tête du dépôt des cartes et plans de la Marine, il ne pouvait qu'être content de connaître ce les Anglais font dans un genre analogue au sien[...] J'achève le premier mémoire de la relation de mon voyage. Ce mémoire, qui comprend tout ce qui est relatif à l'artillerie, sera le plus propre à intéresser le ministre comme général d'artillerie[...] Je viens de recevoir les cartes et les livres que j'ai achetés en Angleterre pour fournir des matériaux à mon travail. La douane de Dunkerque, en faveur de la 'science des a + b', a bien voulu ne pas me prendre de droits, ce qui eût achevé de me ruiner. Je lance dimanche prochain ma corvette avec toute la solennité requise en pareil cas. Je n'ai voulu clore cette lettre qu'aujourd'hui, pour t'annoncer que j'ai lancé ma corvette avec un plein succès. Je

joins ici la note de cette cérémonie vraiment imposante et tu peux la faire insérer au "Moniteur".
M. Sané vient de m'écrire que le ministre, reconnaissant l'utilité et l'importance des résultats de mon voyage, m'a accordé un élève du Génie maritime auquel il va faire faire 300 lieues pour m'aider à dessiner les plans qui vont accompagner mes mémoires [...] Je vais jouir d'un peu de tranquilité maintenant que ma corvette est lancée, et cela tournera tout au profit du récit de ce que j'au vu en Angleterre. (DRB).

54. Letter from Dupin to his elder brother, 27 February 1817 (DRB):
Cette somme de 800 F que j'ai emportée auparavant, plus mon voyage de Paris à Dunkerque, plus
2 400 constituent les dépenses effectives de mon voyage, sans compter le délabrement de ma
garde-robe.
M. Sané me dit de retourner à Toulon dès que la corvette sera partie, mais il faut d'abord:
1) que je rende mes comptes pour 1816-1817,
2) que j'achève la rédaction de mon voyage,
3) que je retourne pour trois mois en Angleterre (si toutefois on m'indemnise de ma première course),
4) que j'imprime mon voyage
La Société Royale de Londres vient de me faire un honneur très grand; elle imprime dans ses
Transactions philosophiques le mémoire que j'ai rédigé sur la construction des vaisseaux avant
de venir ici.
Je crois que, depuis Franklin, qui était à demi-Anglais, on n'a fait cet honneur à aucun étranger.
Je prouverai par là aux Anglais que je sais reconnaître leur accueil libéral et, aux Français, qu'on fait quelque cas de mes travaux à l'étranger.

55. Dupin to his elder brother, 2 March 1817, from Dunkirk (DRB):
Je te prie de recevoir le Dr Young, secrétaire de la Société royale de Londres et l'un des savans les plus distingués de la capitale de la

Grande-Bretagne. Il vient à Paris. C'est un des savants en qui j'ai trouvé le plus d'obligeance et qui m'a rendu des services essentiels.

56. Dupin to his elder brother, 22 March 1817 (DRB):
M. Sané voudrait que je fusse déjà à Toulon. Quand je suis à Paris, il brûle de m'exiler en province et, quand il m'a poussé au nord, il me repousse au midi.
Tâche de modérer un peu la "fureur déplaçante" de M.. Sané. Puisque le ministre m'a envoyé un officier pour m'aider, il est évident qu'il a voulu que j'achevasse la relation de mon voyage, qui va former un énorme in-4°, dont je lui ai déjà donné le premier tiers, dont le deuxième tiers sortira dans dix jours et le dernier dans deux mois, au moment où je voudrais retourner à Londres pour compléter beaucoup de choses importantes.
57. Dupin to Jacques Charles, 18 April 1817 (ADN):
Enhardi par l'indulgence avec laquelle l'Institut a reçu mes travaux, particulièrement sous votre présidence, j'ose me mettre sur les rangs pour remplacer M.Rochon dans la section de physique.
Je joins à cette lettre une notice de mes travaux; puissent-ils mériter votre honorable suffrage. Je vous respecte trop pour rien ajouter à ce que ces travaux peuvent dire en ma faveur; je dois m'en rapporter avec une entière confiance à votre impartialité.
P.S. Oserai-je vous prier de faire remettre les lettres ci-jointes à vos collègues auxquels elles s'adressent?
58. Dupin to Dupin *aîné*, Dunkirk, 15 April 1817 (DRB):
Je demande à être mis sur la liste des candidats parmi lesquels on doit choisir à l'Institut la succession du physicien Rochon qui vient de mourir. Je t'adresse une lettre de M. Lefebvre-Joneau avec une notice de mes titres que je te prie de lire. Si tu pouvais aller voir M. Lefebvre, ce serait très important por moi: il est l'un des cinq membres de la section qui doit composer et présenter la liste des candidats. Les autres sont Poisson, Gay-Lussac et Charles.
Recommande-moi à M.Dayeux (montre-lui la notice que je t'envoie pour en faire le meilleur usage possible) et à d'autres membres de l'Académie des sciences, si tu en connais d'autres. Tu sais que le

sort de ma vie dépend de mon entrée à l'Institut: tout, alors, prendrait une autre face dans mes affaires.
Quand même je ne serais pas reçu, il n'en serait pas moins pour moi d'une extrême importance que
je fusse présenté et que j'eusse un certain nombre de voix: cela prouverait qu'à la première présentation, je serai reçu d'emblée.
P.S. J'ai envoyé au maréchal Marmont, académicien libre, la table analytique de mon mémoire sur l'artillerie anglaise. Il m'a répondu: "L'ouvrage que vous m'annonciez doit être d'une haute importance; si vous avez la complaisance de me le communiquer, je me chargerai bien volontiers de le présenter à l'Académie".

59. Dupin to Dupin *aîné* 2 May 1817 (DRB):
Le ministre a décidé que je retournerai en Angleterre. J'attends impatiemment sa permission. Je voudrais partir du 25 au 30 de ce mois: il est important que j'arrive à temps pour ne pas trouver, comme l'an passé, tous les Anglais au diable et loin de Londres. M. Sané paraît bien disposé pour moi: j'ai reçu de M. Rolland, inspecteur général adjoint, une lettre très flatteuse pour mon travail relatif à l'artillerie.
P.S. Comme abondance de biens ne nuit pas, si tu trouves encore quelque bonne recommandation pour Londres, envoie la moi.

60. Dupin to Bajot from Dunkirk, 19 May 1817 (ADN):
Mon cher Bajot
Je vous adresse une notice sur la mesure trgonométrique tendant au relèvement trigonométrique des côtes de Grande-Bretagne, pour l'insérer dans vos annales. Quand je serai à Londres, je crois que vous ferez bien de m'adresser deux ou trois exemplaires du numéro où sera cet article. J'intéresserai par là directement à votre ouvrage le gouverneur de l'école de Woolwich, qui a paru fort flatté de cette notice qui le concerne particulièrement. Vous savez que l'intérêt personnel est le grand mobile de la pauvre humanité et les gens de lettres estiment beaucoup les écrits périodiques où l'on parle bien d'eux. Ils voudraient les voir universellement répandus.
Au moment où j'écris, on m'apporte votre lettre qui me verse du baume dans le sang. Dès que mes papiers seront expédiés, je partirai.

Continuez-moi votre bonne amitié, mon cher Bajot, et disposez de moi en France comme en Angleterre.
Je vous embrasse de tout mon cœur. Ch. Dupin.
P.S. Madame Courcier va vous faire passer un paquet pour moi à l'adresse de M. Séguier consul général de France à Londres. Pourrez-vous m'obliger en faisant passer ce paquet par le ministère des affaires étrangères; alors, le courrier de l'ambassadeur le porterait. Si cela souffrait quelque difficulté, je vous prierais d' adresser ce paquet purement et simplement à M. Jacques commissare de marine à Calais sous le couvert du ministère de la Marine. Il s'agit de choses qui importent au succès de mes recherches et vous y prenez un intérêt si amical que je ne dois pas craindre de fatiguer votre complaisance.

61. Dupin to his elder brother, 19 May 1817 (DRB):
Je pars lundi prochain pour Londres avec les 1200 F qu'on me donne pour les 6 mois de mon premier voyage et un trimestre qu'on me paie d'avance. J'aurai, avec quelques autres économies, de quoi faire mon voyage. Ainsi, je te remercie pas moins de tout cœur pour la lettre de crédit que tu m'as fait passer.

62. Dupin to Bajot, London 2 June 1817 (ADN):
Mon cher Bajot,
J'ai déjà trouvé l'occasion de faire quelque chose qui, je pense, pourra vous être agréable. En rendant visite à M. Brand secrétaire de la Société royale de Londres et rédacteur du Journal des Sciences et des Arts (of Scieces and the Arts). Je lui ai offert un échange entre vos Annales[1] et son ouvrage périodique Je lui ai fait l'éloge de votre écrit périodique avec la même assurance que si je n'y mettais ordinairement aucun article et je l'ai décidé à accepter le troc proposé. Le journal des Sciences et Arts est conçu sur un très bon plan; les travaux des Sociétés royales de Londres et d'Edimbourg, de l'Institut et des principales académies de l'Europe, et l'analyse des ouvrages nouveaux s'y trouvent. Ce qu'il y a d'heureux, c'est que ce journal a commencé comme le vôtre en

[1] Les annales maritimes

1816. Nous sommes convenus que vous m'enverrez votre collection et que je recevrai pour vous, en échange, celle de M. Brand. Je vous conseille d'envoyer 1816 et 1817 par numéros séparés; cela paraîtra davantage et aura plus de prix.

63. From Dupin in London to his elder brother, 13 June 1817 (DRB):
Je viens d'adresser au ministre, à MM. Sané, Laplace etc., mon mémoire imprimé des Transactions philosphiques de la Sociéte royale de Londres.
J'en ai fait tirer 25 exemplaires qui m'ont coûté 80 francs; c'est le tarif de l'Arabie!
L'ambassaeur et M. Séguier m'ont parfaitement reçu. J'ai pareillement à me louer des Anglais que j'ai revus. J'ai déjà visité deux arseaux. Je partirai le 15 juillet pour l'Ecosse.

64. From Dupin to Bajot, London 26 June 1817 (ADN):
Mon cher ami
On m'a prié de faire l'analyse de vos Annales pour le Journal des sciences et des arts de l'Institution royale de Londres. J'ai travaillé sans relâche et j'ai rendu compte dans un premier article de tous vos numéros de 1816. On a de suite fourré cet article dans le numéro qui va paraître dans cinq jours et j'ai corrigé ce matin l'épreuve qu'on m'a apportée.
Je vous conseille de mettre l'article suivant dans votre numéro qui va paraître. Le Journal des sciences et des arts vient de rendre sur les numéros des Annales maritimes publiées en 1816, un compte rendu très détaillé que nous donnerons textuellement dans notre prochain numéro. Nous répondrons à cette flatteuse marque d'estime qui nous est donnée par le premier journal scientifique de l'Angleterre, en faisant connaître avec soin toute la partie des ses matières qui peut avoir quelque intérêt pour la marine.
Voici un petit article que je vous prie d'insérer dans une page différente du précédent: "La Société royale de Londres, à laquelle M. Dupin a présenté un travail très étendu sur la Structure des vaisseaux anglais, a jugé ce travail digne de paraître dans la collection des ses mémoires. En conséquence, quoique cet écrit soit en langue française, elle vient de le publier dans la première partie du volume

de ses Transactions philosophiques pour 1817. Nous donnerons des extraits de cette nouvelle production de M. Dupin".

Adieu, mon cher ami, je vous embrasse de tout mon cœur.

65. Dupin to Dupin *aîné*, London 11 July 1817 (DRB):

J'ai été accueilli dans la maison du Dr Somerville avec une hospitalité trop rare et dont on ne sent pas le prix que quand on a longtemps voyagé. M. le Dr Somerville, inspecteur des hôpitaux militaires et membre de la Société royale de Londres n'est pas moins distingué par la bonté de son caractère que son talent, et Madame Somerville, son épouse, est célèbre par ses vastes connaissances dans les lettres et les sciences y compris le grec et les mathématiques.

Ainsi, mon cher frère, j'espère que tu traiteras M. et Mme Somerville comme je traiterais les personnes que tu me recommanderais le plus particulièrement.

Dans huit jours, je reprends la route du Juif errant. Je vais d'abord à Edimbourg.

66. Dupin à Dupin *aîné* , London 15 July 1817 (DRB):

Je vais partir pour l'Ecosse sans avoir de tes nouvelles et de celles de ma famille.

Le Dr Somerville, qui part demain pour la France, viendra te remettre deux lettres pour toi, ta femme et Philippe.

Il a eu des bontés pour moi et m'a procuré d'excellentes recommandations pour l'Ecosse; aussi, je te prie de lui faire honnêteté ainsi qu'à Madame Somerville.

67. Dupin to Bajot from London, 20 July 1817 (ADN):

Lisez ma lettre à M. Portier pour vous mettre au fait de mes courses et appuyer ma demande. Je pars après demain. J'ai reçu vos deux numéros de juin. J'ai remis de suite l'un d'eux au rédacteur du Journal des sciences et arts qui a paru flatté de notre petite note. Il faut que vous ayez la patience d'attendre mon retour à Londres, où je m'empresserai de vous faire un article détaillé au sujet de ce journal. Je désire que mon mémoire sur les vaisseaux anglais produise plus d'effet qu'il n'en a produit à Paris, manuscrit.

Je m'en vais avec environ quarante letters de recommandations pour des savans, littérateurs, ingénieurs et autres; il faut tout cela pour bien voir. Je vais économiser ma petite bourse et tant qu'il me restera un sol, j'irai en avant. Je ne reviendrai en France que quand il ne me restera plus rien des sept mille francs que j'ai consacrés à ces deux voyages.

68. Dupin to Bajot, 15 August 1817, from Edinburgh (ADN):
Je vais partir d'Edimbourg dans quelques jours pour aller voir le phare de Bell-Rock, la fonderie de Carrow, etc. Il fait dans ces pays un tems épouvantable et c'est dommage, car il y a des choses bien intéressantes à voir.
Si l'on savait la peine et la dépense que l'on éprouve et que l'on fait dans ce pays, on ferait quelque chose pour ceux qui sacrifient leur fortune et leur santé à leur patrie.
69. Dupin to the Minister for the navy, from Liverpool, 7 October 1817 (ASHM):
Avant de quitter l'Angleterre pour compléter mon "Tableau de l'architecture navale aux 18e et 19e siècles", et pour étudier en même temps tous les autres sujets relatifs à la Marine, je crois devoir proposer à votre Excellence l'achat de quelques objets dont l'acquisition me paraît d'une importance majeure. Il s'agit d'acheter des câbles en fer battu, et des garnitures d'écubier, des palans en fer battu destinés aux grands manœuvres de force, des scies circulaires, des verres optiques rouges tels que je viens de voir employés au célèbre phare de Bell-Rock sur la côte d'Ecosse'.
70. Dupin to Dupin *aîné*, 22 November 1817 (DRB):
En arrivant à Dublin, je me suis donné une solide entorse. J'ai quitté mes béquilles, je marche assez joliment et, dans peu de jours, je bats en retraite pour Londres et la France.
71. Dupin to Bajot, 22 November 1822 (ADN):
Je vais exposer les causes de mon long silence. Après avoir couru sans accident majeur le montagnes et les précipices de la Haute-Ecosse, je suis venu me rompre les talons à Dublin, en y rentrant à minuit et tombant de cinq pieds de hauteur seulement. Il y a 36 jours que cela

m'est arrivé; j'ai déjà quitté mes béquilles et je cloche à présent fort joliment en m'appuyant sur un seul bâton.

72. From Séguier to the Minister for the Navy and the Colonies, 8 February 1818 (ASHM):
    Charles Dupin a évité toutes les dépenses inutiles; au bout de six mois, alors qu'il était arrivé ignorant de la langue anglaise, il était capable de la parler très bien, ce qui lui ouvrit toutes les portes. Les Anglais sont en général peu communicatifs et très réservés sur le secret de leur industrie.
73. Dupin to Navier, 11 June 1819, from Dunkirk (ADN):
    Je viens de recevoir un exemplaire des "Mémoires sur la Marine et les Ponts-et-Chaussées" que je viens de publier. Je désire que cet ouvrage obtienne votre suffrage. Vous pourriez me rendre un grand service que j'ose réclamer de vous comme d'un ancien camarade d'Ecole et qu'en pareille occasion je vous rendrais de tout mon coeur si j'en étais capable. Ce serait de faire, sur la partie "Travaux maritimes et ponts" de cet ouvrage, une analyse [...]; nous la ferions paraître dans le Moniteur et les ingénieurs de votre Corps, en voyant votre suffrage qui fixerait leur confiance perdraient les préjugés qu'ils prendront naturellement d'un constructeur qui parle bassins, ponts et canaux. [...] J'ai dédié mon ouvrage à M. de Prony, notre ami commun; ce doit être pour vous un motif de plus pour prendre intérêt à cette collection.
74. Dupin to the British Museum librarian, Dunkirk 20 June 1818 (ADN):
    L'Ingénieur de la Marine en chef
    J'ai l'honneur de vous adresser trois exemplaires de mas "Mémoires sur la Marine et les Ponts-et-Chaussées", le 1er pour vous, le second pour Dr Brewster à Edimbourg; le 3ème pour M. Jeffries, avocat à Edimbourg et rédacteur de l'Edimbourg Review.
    J'attends de votre complaisance que vous voudrez bien remettre ces exemplaires à leur adresse. Ils vous donneront d'ailleurs occasion de connaître des hommes d'un rare mérite et d'une haute réputation.
75. Le Prince de Polignac au comte de Chabrol, ministre secrétaire d'Etat au département de la Marine,
    London 1824 (ASHM):

Ce que je puis assurer, c'est que le travail immense auquel il s'est livré est tellement apprécié en Angleterre que les membres du ministère anglais m'ont dit plusieurs fois y avoir puisé des informations qu'ils ne pouvaient trouver ailleurs.

76. Dupin *père* to his second son, 26 March 1819 (DRB):
Je m'empresse de me féliciter avec toi de ton nouveau grade. C'est pour ma famille un nouveau motif d'attachement au roi. C'est lui qui t'a fait membre de la Légion d'honneur, de l'Institut, et officier supérieur du Génie maritime. Je vois que tu as eu soin de consulter plusieurs gens de lettres pour te faire indiquer les fautes de plan, de pensées, de style, etc. Mais il ne suffit pas de demander des avis, il faut les suivre; il ne convient pas qu'un académicien soit tout à la fois hérétique et schismatique en frondant, par son exemple, l'autorité et l'Académie française en ce qui concerne la langue.

77. Dupin *père* to his son Charles, 11 January 1821 (DRB):
De tous les arts et métiers enseignés au Conservatoire, je n'en connais pas de plus intéressant, mon ami, que l'art épistolaire ou, si l'on veut, le métier d'écrire, quand c'est entre gens qui en valent la peine, par exemple des pères aux enfans.
Tu m'as souhaité une bonne année. Je te la souhaite aussi et, sous ce rapport, nous voilà quittes! [....]
Ce n'est pas moi qui te presserai, car je connais trop peu de femmes dont je voulûsse être le mari et trop peu d'enfans dant je désirasse être le père. En attendant, fais tes cours, tes courses et ta bourse.

78. Dupin to his elder brother, 1 September 1821 (DRB):
Le ministre de la Marine est bien disposé pour moi, mais le ministre est un homme qui ne peut se décider à rien; quand les Bureaux lui ont demandé de me faire offcier de la Légion d'honneur, il n'a répondu oui ou non.
Si tu pouvais écrire à Roy, cela me serait fort utile. Le 13 septembre, anniversaire du baptême du duc de Bordeaux, sera sans doute une époque où l'on donnera de nouvelles décorations.
Le Moniteur, le Constitutionnel et le Courrier ont rendu compte de la "Force navale" dans des articles étendus et substanciels [sic].

Pendant ce mois d'août, j'ai déjà préparé beaucoup de matériaux pour la troisième partie de mes voyages et beaucoup de choses particulièrement utiles pour notre partrie, car c'est là le but principal de mes travaux.

79. Dupin to the Ministre de la Marine et des Colonies, Paris, 9 October 1820 (ASHM):
    Monseigneur,
    a) Prochainement, je commencerai mes leçons de "méchanique appliquée aux arts", prescrites par une Ordonnance de Sa Majesté.
    b) Mon objet le plus cher: que ce cours soit utile aux arts de la Marine.
    c) Or, cette année, Polytechnique va fournir plusieurs élèves au Corps du Génie maritime.
    d) Je propose donc que ces jeunes gens restent l'hiver à Paris pour y suivre mes cours, avant d'être envoyés dans les ports.
80. From Rolland, *inspecteur général du Génie maritime*, to the *Ministre de la Marine et des Colonies*,
    23 October 1820 (extract, ASHM):
    a) M. Dupin, ingénieur de la Marine, professeur pour le cours de mécanique au Conservatoire des Arts et Métiers a proposé de faire suivre son cours aux élèves sortant de Polytechnique et affectés au Génie maritime.
    b) Voici mes observations:
       Si ces élèves restaient ainsi un an à Paris, ce cours serait leur seule occupation.
    c) Or, les familles de ces jeunes gens pourraient craindre que leurs enfants "livrés à eux-mêmes dans l'âge des passions ne se livrent à des habitudes contraires à la morale".
    d) Ce risque ne serait pas à craindre dans les ports, car "les occasions de se livrer au vice sont moindres qu'à Paris".
81. Dupin's reply in a letter to the *Ministre de la Marine*, 4 November 1820 (ASHM):
    a) Les élèves de Polytechnique, affectés au Génie maritime doivent recevoir un enseignement pratique en plus de leur formation technique. D'où l'utilité de suivre mon cours de Conservatoire.

b) La jeunesse n'a certainement pas moins de tentations dans les port qu'à Paris, "pour le jeu, le vin et les femmes".

82. Dupin to 'M. le Duc', 11 February 1831 (Archives of the CNAM):
Avant-hier seulement, j'ai reçu le rapport sur lequel le Conseil du Conservatoire doit aujourd'hui délibérer.
Je ne puis concevoir qu'on propose à ce Conseil, protecteur naturel des Ecoles d'Arts-et-Métiers, de détruire les deux Ecoles de Châlons et d'Angers; M. de Corbière même n'aurait pas osé suivre un tel conseil.
Les reproches qu'on fait à ces établissements sont exagérés. On peut aisément remédier aux défauts qu'on leur reproche. Comme la mémoire du Duc de Rochefoucault-Liancourt, ces établissements sont liés à la gloire nationale; il y aurait du vandalisme à les détruire.
Je me borne à déposer entre vos mains ma protestation contre la suppression des Ecoles de Châlons et d'Angers; je plaiderai leur cause auprès du Ministère et, s'il le faut, dans les Chambres.

83. Dupin to the Duc de la Rochefoucauld, 11 September 1825 (ADN):
Si vous pouviez m'envoyer un jeune homme d'espérance pendant neuf à dix semaines seulement, par exemple du 15 octobre à la fin de décembre, ce temps lui suffirait pour suivre mon cours de géométrie des arts et métiers, apprendre un peu le dessin linéaire et revenir au 1er de l'an à Liancourt donner le soir des leçons de géométrie à vos ouvriers. [...] Je ferais si bien occuper ce jeune homme qu'il n'aurait pas le temps de se débaucher. Ces leçons ensuite pourraient être écoutées avec une égale facilité par les jeunes et les vieux [...]
Je tiens beaucoup, je l'avouerai, pour l'honneur de l'industrie française, que ce cours soit fait pour les ouvriers du Duc de la Rochefoucault, le créateur des écoles publiques d'Arts-et-Métiers: c'est à vous de compléter cette partie de votre gloire.

84. Dupin to his daugher, Marie, from London, 17 May 1851 (ADN):
Je profite de l'occasion offerte par le départ d'un Français pour te remercier de ta charmante lettre; elle est écrite en très bon anglais, sans faute d'orthographe ni de grammaire, et fort bien pensée. [...]

J'ai été invité à un dîner public, où nous sommes restés à table cinq heures et demie. Nous avons digéré, plus ou moins, trente-deux discours, y compris le mien que j'ai fait en anglais, et qu'on a reçu avec beaucoup de complaisance. Je souhaite bien que Charles travaille. Faites-lui la morale, ce mot qu'il abhorre!

86. Dupin *père* to Charles, from Clamecy, 10 July 1834:
'A Ch. Dupin, membre du Conseil général, à Nevers.
Tu vas, mon ami, te trouver au Conseil général avec ton frère. C'est pour moi l'occasion d'un conseil paternel et amical. Concerte toi dans le particulier avec lui, afin que la différence d'opinion sur quelque objet que ce soit, ne puisse engager entre vous deux aucune collision, toujours désagréable en présence de tiers.
Soyez economes des deniers publics.
Favoriser la religion, les bonnes mœurs, l'instruction convenable aux différents états ou professions, le bon choix des fonctionnaires publics, la répression des mauvaises têtes et des tapageurs [...].

87. Letter to Dupin *aîné* from de Rheville, 21 November 1843 (ADN):
Un comité provisoire s'est formé dans le but d'arriver à faire établir un chemin de fer d'Auxerre à Nevers, en passant par Clamecy; il a fait part de ses intentions à M. le baron Charles Dupin, votre frère, qui a bien voulu accepter la présidence d'honneur qui lui a été spontanément offerte.
(Signé) De Rheville.
Siège du Comité: Rue des Filles Saint-Thomas, à Paris.

88. From Dupin to a M. Vortier, 15 November 1834 (ADN):
Monsieur
Vos félicitations, dont je vous remercie de tout mon coeur, me sont arrivées lorsqu'un devoir de bon citoyen et d'homme d'honneur m'avait imposé le devoir d'offrir au Roi ma démission. Je regrette le bien que j'avais l'espoir de faire à la Marine.

89. Dupin to the Duc de Montpensier, 3 July 1847 (ADN):
Permettez-moi d'offrir à votre Altesse royale mon rapport sur le travail des enfants dans toutes les manufactures et les ateliers du royaume. L'avenir du peuple est dans le bien que nous voulons

produire par l'instruction des enfants, dont nous ménageons la force en modérant le travail.

90. Dupin *père* to Charles, 15 May 1838 (DRB):
Ton frère mon ami s'est très cordialement hâté de me donner l'avis officieux de ta nouvelle nomination au grade supérieur de Directeur das constructions navales, mais ma satisfaction à cet égard est tempérée par la crainte que le cumul de tant d'occupations diverses ne compromette ta santé.
Autrefois, j'ai dû pousser à la roue du travail mes jeunes fils, mais, à leur âge actuel, je dois aujourd'hui les exhorter à "enrayer" pour ne pas se laisser entraîner au tombeau.

91. Dupin to an unnamed general, 29 July 1862 (ADN):
Dans cette introduction, j'ai voulu résumer, en traits généraux et rapides, les intérêts caractéristiques de l'étendue des forces et le génie spécial des nations que j'ai déjà décrites.
Un lecteur attentif trouvera dans mes recherches la clef de bien des événements; il y trouvera la mesure de bien des succès qui surprennent chaque jour et qui surprendraient si l'on cherchait dans le progrès inégal des nations, depuis soixante ans, l'action, inégale aussi, qu'elles exercent aujourd'hui et qu'elles exerceront dans la seconde moitié du siècle.
J'ai déjà décrit les forces productives de 9 millions d'hommes; je n'en ai plus que 450 millions à développer [...] voilà la principale et dernière occupation de ma vie.

92. Dupin to Rapetti, 8 September 1866 (ADN):
Je poursuis mes travaux sur la partie de l'Orient qui n'est pas en question sur l'Inde; j'en suis à l'Introduction de ma dernière partie et j'ai des [illegible] magnifiques, je ne dirais pas à développer mais à condenser'.

93. Dupin to Rapetti, 1 September 1871 (ADN):
Vous pouvez juger avec quelle anxiété patriotique et douloureuse j'étudie les phâses des propositions et de la délibération sur le fonctionnement du pouvoir constituant et la Présidence [....] Quel sera l'avenir, trop prochain peut-être, de ce système sans exemple? Je

ne désire pas prévoir et j'attends [...] J'ai trouvé Gambetta sans habileté, sans adresse et violent. Gardons-nous de nous en plaindre. [...]
Et plus que jamais, dans ma solitude,
Je respire la paix et l'indépendance dont
Le sage doit se faire un bonheur.

Chronology

Dupin's Life and Work

7 October 1784: Born at Varzy (Nièvre).

1801: Admitted to the Ecole polytechnique.

1802: Attracts the attention of Monge and Carnot for his mathematical discoveries.

1803: Graduates from the Ecole polytechnique to join the naval engineering corps; stationed at the Boulogne arsenal.

1804: Transferred to the Antwerp arsenal.

1805: Transferred to the Genoa arsenal.

1807: Transferred to the Toulon arsenal.

1808: Joins the Corfu arsenal.

1808: Participates in the creation of the Ionian Academy.

1811: Leaves Corfu because of illness; has a year-and-a-half-long convalescence in Italy (Naples, Rome, Florence, and Pisa).

1812: Returns to Paris and receives praise from the first class of the Institut for his work on the geometry of curved surfaces.

1813: Returns to the Toulon arsenal and becomes a corresponding member of the Institut.

1816: Sent to the Dunkirk arsenal.

July 1816–February 1817: Makes his first visit to Britain.

June 1817–February 1818: Makes his second visit to Britain, followed by other study visits in 1819, 1821, and 1824. He would make a sixth (and final) visit in 1851 in connection with the Great Universal Exhibition in London.

1818: Enters the Academy of Sciences.

1819: Takes the chair of Géométrie et mécanique appliquées aux arts, at the Conservatoire des arts et metiers, giving his first lecture on 11 November 1820. He teaches there until 1865.

1823: Becomes chevalier de l'ordre de Saint-Louis.

1824: Made a baron by Louis XVIII and an officer of the Légion d'honneur.

1827: Elected député for the Tarn (until 1830).

1830: Marries Rosalie-Anne Joubert (1806–1876); they will have three children. Elected député for the Seine.

1831: Appointed conseiller d'Etat en service extraordinaire (until 1848).

1832: Active service in the corps du Génie maritime comes to an end. He is co-opted to the Académie des sciences morales et politiques in the political economics and statistics section. Appointed commandeur de la Légion d'honneur.

1834: Appointed minister of the navy and colonies (from 10 to 18 November).

1837: Receives the title of pair de France.

1838: Receives an honorary promotion to the grade of director of naval construction.

1840: Named a grand officier de la Légion d'honneur.

1843: Receives an honorary promotion to the grade of inspecteur général du Génie maritime.

1848: Elected representative for the Seine-Inférieure at the Constituent Assembly; reelected to the Legislative Assembly in 1849 (until 1851).

1851: Appointed chairman of the French committee that is sent by the minister of trade to the Great Universal Exhibition in London.

1852: Leaves the Corps du génie maritime. Created senator by Napoleon III.

1870: Starts to leave public life.

18 January 1873: Dies in Paris, to be interred in the family vault at Clamecy.

Primary Bibliography

Works by Charles Dupin

Dupin was such a prolific writer and speaker that it is impossible to produce a fully comprehensive list of his works. Grattan-Guinness did sterling work in that direction, and I have been able to draw extensively his bibliography. In his *Convolutions in French Mathematics, 1800–1840* (1990, 1107–1110, 1463–1465), he provided meticulous details of Dupin's manuscripts as well as published reviews, summaries, and reports relating to many of the works that are mentioned in this book.

1807a. Essai sur la description des lignes et des surfaces du second degré. *Journal de l'Ecole polytechnique* 1, no. 7, *cahier* 14:45–83. Paris: Imprimerie impériale.

1807b. Analyse d'un mémoire sur la théorie des déblais et remblais (1807). *Correspondance sur l'Ecole polytechnique* 1 (1804–1808): 118–125. Paris: Hachette.

1809. Sur l'éducation publique des Grecs. Address delivered at a public meeting of the Ionian Academy, 15 August. DRB.

1812. *Examen des travaux de César au siège d'Alexia, oeuvre posthume de Léopold Vacca Berlinghieri, avec la vie de cet auteur*. Paris: Bachelier.

1813a. *Expériences sur la flexibilité, la force et l'élasticité des bois, avec des applications aux constructions en général, et spécialement à la construction des vaisseaux, faites à l'arsenal de la marine française à Corcyre en 1811*. Nîmes: Blachier-Belle.

1813b. *Développements de géométrie, avec des Applications à la stabilité des Vaisseaux, aux Déblais et Remblais, au Défilement, à l'Optique, etc pour faire suite à la Géométrie Descriptive et à la Géométrie analytique de M. Monge*. Paris: Vve. Courcier.

1814a. *Essais sur Démosthène et sur son éloquence*. Paris: Vve Courcier. In this work, he produced a translation of the Olynthian orations of Demosthenes. To this, he added *Considérations sur l'éloquence de l'auteur athénien*.

1814b. *Des lois fondamentales de la France, au sujet de la constitution de 1814*. Toulon: A. Aurel.

1815a. *Du rétablissement de l'Académie de marine*. Paris: Vve Courcier.

1815b. Expériences sur la flexibilité, la force et l'élasticité des bois. *Journal de l'Ecole polytechnique* 1, no. 10, *cahier* 17:137–211.

1815c. *Tableau de l'architecture navale militaire aux XVIIIe et XIXe siècles*. Paris: Vve Courcier. An analysis of the first part of this work was presented on 12 June 1815 to the first class of the Institut de France.

1817. Rapport au ministre. 19 May 1817, ASHM/VI/BB1 53, fol. 102.

1818a. Aux moyens de mettre en harmonie la branche la plus important du pouvoir public avec la forme et l'opération d'un gouvernement représentatif et monarchiste. ENPC, ms. 3000.

1818b. *Mémoires sur la Marine et les Ponts-et-Chaussées de France et d'Angleterre*. Paris: Bachelier.

1818c. *Réponse au discours de Milord Stanhope sur l'occupation de la France par l'armée étrangère*. 1st ed. London: Barnes and Noble. 2nd ed. Paris: L'Huillier.

1818d. *Travaux théoriques et pratiques de Charles Dupin*. Paris: Jeunehomme-Crémière.

1818e. *Lettre à Mylady Morgan sur Racine et Shakespeare*. Paris: Bachelier.

1819a. *Two excursions to the ports of England, Scotland and Ireland, in 1816, 1817, and 1818*. London: Sir Richard Phillips and Co.

1819b. Note au second voyage dans les ports d'Angleterre d'Ecosse et d'Irlande, par M. Charles Dupin. ASHM/V1/CC7 779.

1819c. *Essai historique sur les services et les travaux scientifiques de Gaspard Monge*. Paris: Bachelier.

1819d. L'influence des sciences sur l'humanité des peoples. Lecture to the Institut, 1819, later published in Dupin (1825c, 1, 63–88).

1820a. *Force militaire de la Grande-Bretagne*. 2 vols. and atlas. Paris: Bachelier.

1820b. Progrès des sciences et des arts de la marine française depuis la paix. Lecture to the Académie des sciences, *Procès verbaux* 7, no. 34. Later published in Dupin (1825c, 1, 89–132).

1820c. Introduction au cours de géométrie et de mécanique; la force ouvrière. Opening lecture at the Conservatoire des arts et métiers. Later published in Dupin (1825c, 2, 1–43).

1821a. Discours d'inauguration de l'amphithéâtre du Conservatoire des Arts et Métiers (8 January 1821). In *Annales de l'industrie nationale et étrangère ou Mercure technologique, recueil de mémoires sur les Arts et Métiers*. Paris: Bachelier.

1821b. *Force navale de la Grande-Bretagne*. 2 vols. and atlas. Paris: Bachelier.

1821c. Considérations de l'industrie et des machines en France et en Angleterre. Lecture to the Institut. Later published in Dupin (1825c, 1, 142–173).

1822a. *Applications de géométrie et de mécanique, à la marine, aux ponts et chaussées, etc., pour faire suite aux Développements de géométrie*. Paris: Bachelier.

1822b. *Influence du commerce sur le savoir, sur la civilisation des peuples anciens et sur leur force navale*. Paris: Bachelier. Also in *Annales de l'industrie nationale*, 6, no. 22: 178–195, 215–241. Later published in Dupin (1825c, 1, 228–276).

1822c. Sur la stabilité des corps flottants. Published in Dupin (1822a, 1–74).

1822d. Du tracé des routes isolées. Published in Dupin (1822a, 75–126).

1822e. Sur le tracé des routes dans les déblais et les remblais. Published in Dupin (1822a, 127–186).

1822f. Sur les routes suivies par la lumière et par les corps élastiques. Published in Dupin (1822a, 187–245).

1822g. Inauguration of the CAM amphitheatre. *Annales de l'industrie nationale* 5, no. 22:5–30. Later published in Dupin (1825c, 2, 44–70).

1823a. *Du commerce et ses travaux publics, en Angleterre et en France*. Paris: Bachelier. Later published in Dupin (1825c, 1, 281–326).

1823b. *Système de l'administration britannique en 1822, considéré sous les rapports des finances, de l'industrie, du commerce et de la navigation, d'après un exposé ministériel*. Paris: Bachelier.

1823c. *Rapport sur les avantages, sur les inconvénients et sur les dangers comparés des machines à vapeur*. Paris: Bachelier. Also in *Annales de l'industrie manufacturière* 10:113–152; PV 7, 469–479.

1823d. Rapport sur les ponts de Navier. *Moniteur Universel*, 1221–1222, 1302–1302, 1317–1318. Published (Agasse) = *Procès verbaux de l'Académie des sciences* 7:549–559 (Extracts in the MU and the complete paper in the Procès-verbaux the regular report of the Academy proceedings).

1823e. Progrès de l'industrie française depuis 1800, opening lecture at the CAM. Later published in Dupin (1825c, 2, 71–123).

1823f. L'industrie dans le département de la Moselle. In *Revue encyclopédique*, 19, 23, 517–525.

1824a. *Des sociétés bibliques formées par des femmes*. Paris: Sétier.

1824b. *Observations sur la puissance de l'Angleterre et sur celle de la Russie*. Paris: Bachelier.

1824c. *Progrès de l'industrie française depuis le commencement du XIXe siècle*. Paris: Bachelier.

1824d. Avantages sociaux d'un enseignement public appliqué à l'industrie. *Annales de l'industrie nationale et étrangère* 15, no. 56:113–135. Later published in Dupin (1825c, 2, 124–148).

1824e. *Force commerciale de la Grande-Bretagne*. 2 vols. and atlas. Paris: Bachelier.

1824f. Introduction d'un nouveau cours de géométrie et de mécanique au conservatoire. *Annales des sciences et de l'industrie* 16:193–216, 225–252. Later published in Dupin (1825d, 2, 149–209).

1824g. Programme d'un cours de géométrie et de méchanique au Conservatoire. *Annales de l'industrie nationale et étrangère* 16 (1824): 181–192, 296–311; 17 (1825): 88–97, 206–211, 290–314; 18 (1825): 90–97. Published in full in Dupin (1825c, 2, 245–332).

1825a. Sur la régénération de la Grèce, par les progrès des sciences, de l'industrie, de la marine et du commerce. In *Discours et leçons sur l'industrie, le commerce, la marine et sur les sciences appliquées aux arts*. 2 vols. Paris: Bachelier.

1825b. *Géométrie et méchanique des arts et métiers et des beaux arts*. 3 vols. Paris: Bachelier. This was published as three parts: Géométrie (1825), Mécanique (1826), and Dynamie (1826). It was translated into German (3 vols., Berlin: Universitat die Kunste) and Dutch (vol. 1, Amsterdam: Ten Brinck and de Vries) in 1827. It was also translated into English (1827, London: Tait), Italian (1829, 3 vols., Florence: Nella Stamperia di Guglielmo Piatti), Polish (1827–1828, Warsaw: Druck N. Glucksberg), and Norwegian (1829–1832, 3 vols., Kjobenhavn: Reitzel).

1825c. *Discours et leçons sur l'industrie, le commerce, la marine et sur les sciences appliqués aux arts*. 2 vols. Paris: Bachelier.

1825d. Second discours au Conservatoire. *Annales de l'industrie nationale et étrangère* 17:175–206. Published in Dupin (1825c, 2, 245–332).

1825e. Discours de fermeture au CAM: applications de la mécanique à la vie et à l'industrie. Published in Dupin (1825c, 2, 333–391).

1825f. Notice sur un nouveau cours de géométrie et de mécanique et sur son 1825c. *Procès-verbaux de l'Académie des sciences* 8: 262 and in full in the *Annales de l'industrie nationale*, 19, 25, 299–312.

1825g. *Voyages dans la Grande-Bretagne entrepris relativement aux services publics de la guerre, de la marine et des ponts et chausses en 1816, 1817, 1818, 1819, 1820 et 1821*, Paris, Bachelier, 6 vols.

1826a. *Tableau des arts et métiers et des beaux-arts*. Paris: Bachelier.

1826b. *Effets de l'enseignement populaire, de la lecture, de l'écriture et de l'arithmétique, de la géométrie et de la mécanique, appliquée aux arts, sur les prospérités de la France*. Paris: Bachelier. *Annales des sciences et de l'industrie* 1:145–167, 278–298. Later published in Dupin (1827d, 5, 1–51).

1826c. *Tableau statistique ... de Seine-et-Oise. Discours prononcé pour l'ouverture du cours de géométrie et de mécanique appliquées aux arts, à Versailles* (7 November 1826). Paris: Frain.

1826d. Rapport au ministre de la marine sur les cours de géométrie et de mécanique dans les ports de France. *Journal des sciences militaires* 1, no. 4:26, 122–149. MU, 26, 1397–1398, 1401–1402.

1827a. *Situation progressive des forces de la France depuis 1814*. Paris: Bachelier.

1827b. *Conclusion des recherches statistiques sur les rapports de l'instruction publique avec la moralité des diverses parties de la France*. 2 vols. Paris: Guiraudet.

1827c. *Forces productives et commerciales de la France. Conclusion des recherches statistiques sur les rapports de l'instruction publique avec la moralité des diverses parties de la France*. 2 vols. Paris: Bachelier.

1827d. *Le petit producteur français*. 6 vols. Paris: Bachelier (*Petit tableau des forces productives de la France depuis 1814*; *Le petit producteur*; *Le petit fabricant*; *Le petit commerçant*; *L'ouvrier français*; *L'ouvrière française*).

1827e. Discours sur le sort de la classe ouvrière, dans les divers arrondissements de Paris. CAM lecture. Published in Dupin (1827d, 5, 56–104).

1827f. Discours à l'Athénée sur les forces et le sort de la femme. Published in Dupin (1827d, 6:1–40).

1827g. Rapport sur les cours de Poncelet à Metz sur la méchanique. *Bulletin universel des sciences et de l'industrie; sciences technologiques* 8:214–224. Also in *Procès verbaux de l'Académie des sciences* 8:527–531.

1827h. Recherches statistiques sur les rapports de l'instruction populaire avec la moralité. *Procès verbaux de Académie des sciences* 8:481.

1827i. L'application des beaux arts à l'industrie. MU, 27, 1393–1394.

1827j. *Souscription pour une médaille qui doit être frappée en l'honneur de la liberté civile et religieuse de l'univers, à l'occasion de la mort de G. Canning.*[1] Paris: Bachelier.

1828a. *Tableau comparé de l'instruction populaire avec l'industrie des départements, d'après l'exposition de 1827, présenté dans la seconde séance du cours de géométrie et de mécanique appliquées aux arts, professé pour les ouvriers*. Paris: J. Tastu.

1828b. *Forces électorales à la fin de 1827, et situation progressive des forces de la France depuis* 1814. Paris: Bachelier.

1828c. *Navigation intérieure de la France*. Paris: Didot.

1828d. Rapport sur les dangers sociaux des lotteries. *Journal des connaissances usuelles* 7, no. 28, 41–48.

1829. Discours sur la géométrie et la mécanique dans la classe industrieuse. Opening speech at the conservatoire. MU, 123–124.

1830a. *Inspection du chemin de fer de la Loire*. Paris: Huzard-Courcier.

1830b. *Appel aux ouvriers français pour les engager à ne pas briser les machines productives*. Paris: pamphlet. MU, 30, 911.

1831a. Discours du baron Ch. Dupin, député de la Seine, dans la discussion de la proposition de M. Arago, relative à la destruction des Ecoles de Châlons et d'Angers et à la creation d'écoles nouvelles. Chambre des députés, session of 18 November 1831. Extract from the *Moniteur* of 20 November 1831.

1831b. *Discours sur le sort des ouvriers, considéré dans ses rapports avec l'industrie, la liberté et l'ordre public*. Closing speech of Dupin's course at the Conservatoire des arts et métiers, 19 June 1831. Paris: Bachelier.

1831c. Rapport à la Chambre des députés sur une enquête relative à la situation des routes et canaux. *Moniteur Universel*, 22 February 1831. Also in *Journal du génie civil* 10:321–396.

1832a. *Cause du malaise industriel et commercial de la France, et moyens d'y remédier*. Paris: Paulin.

1832b. *Mémoire sur le transport en France des obélisques de Thèbes*. Paris: Imprimerie royale.

1833a. *Discours sur l'avenir de la classe ouvrière*. Paris: David.

1833b. *Harmonies des intérêts industriels et les intérêts sociaux pour servir d'introduction à l'enseignement du Conservatoire des arts et métiers*. Paris: Bachelier.

1834a. *Essais sur l'organisation progressive de la marine et des colonies*. Paris: Bachelier.

1834b. Compte rendu de son premier mandat de député dans le Xe arrondissement de Paris. ADN.

1835a. *De l'influence de la classe ouvrière sur le progrès de l'industrie*. Opening speech for the course in geometry and mechanics applied to the arts et métiers and the fine arts. Paris: Mme de Lacombe.

1835b. Sur les mathématiques en France depuis 1830. Presidential lecture at the Académie des sciences. *Compte rendu de l'Académie des sciences* 1, no. 35:564–574.

1836a. *Tableau des intérêts de la France, relatifs à la production et au commerce des sucres de canne et de betterave*. Presented at the general session of the councils for agriculture, commerce and manufacture, 19 January. Paris: Mme de Lacombe.

1836b. *Rapport du jury central sur les produits de l'industrie française exposés en 1834*. 3 vols. Paris: Imprimerie royale.

1837a. *La Caisse d'épargne et les ouvriers*. Lesson given at the Conservatoire royal des arts et manufactures, 22 March. Paris: Firmin Didot.

1837b. *Faits et calculs relatifs au projet de loi pour réduire les droits d'entrée du sucre français des colonies et du sucre étranger*. Paris: A. Henry.

1838a. *Défense des intérêts coloniaux confiés au conseil des délégués pendant la législature de 1833 à 1838. Compte-rendu par M. le Baron Charles Dupin*. Paris: Firmin-Didot.

1838b. *La Morale, l'enseignement et l'industrie*. Paris: Firmin-Didot.

1839a. Crise commerciale de 1839, examinée dans ses causes, son étendue, et les moyens d'y mettre un terme. Extract from *Mémorial du commerce et de l'industrie*, 15 April. Paris: Bureaux du Mémorial du commerce et de l'industrie.

1839b. *Historique de l'enseignement industriel et de son influence sur le sort du peuple, de 1819 à 1839*. Opening lecture at the conservatoire. Paris: Bachelier. *Moniteur Universel*, 2150–2152, 2164–2166.

1840a. *Conseils adressés aux ouvriers parisiens*. Paris: Firmin Didot frères.

1840b. *Eloge de M. le Baron de Prony*. MU 624–629. Quoted from *Journal du Génie Civil* 11, no. 2 (1846): 326–351.

1840c. Rapport à la Chambre des pairs sur le travail des enfants [22 February 1840] *Compte rendu de l'Académie des sciences* 10, no. 40: 607–612 (2 parts; Part 2 was published in 1847).

1841a. *Bien-être et concorde des classes du peuple français*. Paris: Firmin Didot.

1841b. *Vie d'un bienfaiteur du peuple, A-P. de la Rochefoucauld, duc de Doudeauville*. Paris: Firmin Didot frères.

1841c. *Le Bienêtre des ouvriers*. Paris: Chassaignon.

1842a. *La Vérité des faits sur les cultures comparées des colonies et de la métropole, et pricipalement sur la question des sucres*. Speech to the assembly of the three general councils for agriculture, manufacturing, and trade, 26 December 1841. Paris: Firmin Didot frères.

1842b. *Progrès moraux de la population parisienne depuis l'établisse ment de sa caisse d'épargne*. Paris: Firmin Didot frères.

1842c. *Observations exposées au Conseil Général d'Agriculture*. Paris: Firmin Didot frères.

Discours au roi pour l'anniversaire de sa fête, le 1er Mai 1842, prononcé par le baron Charles Dupin, en présentant le conseil des délégués des colonies. Paris: Firmin Didot frères. (n.d.).

1842d. *Mémoire adressé par le conseil des délégués des colonies aux ministre du roi, sur la question des sucres et sur la crise imminente dont elle menace les ports et les colonies*. Paris: Firmin Didot frères.

1843a. *Appel au bon sens des départements vinicoles, des départements maritimes et de tous les départements du centre, de l'est, de l'ouest et du midi … sur leurs intérêts agricoles, manufacturies et commerciaux, dans la question des sucres*. Paris: Firmin Didot frères.

1843b. *Carte figurative de la France pour exprimer par départements la proportion territoriale des vignobles et des betteraves à sucre*. Paris: Firmin Didot frères.

1843c. *Eloge de M. le maréchal Moncey, duc de Conegliano: À l'usage des écoles régimentaires de l'armée française*. Paris: Firmin Didot frères.

1844a. *Mémoire sur la situation sociale et politique des colonies françaises en 1844, et sur le système que le gouvernement doit adopter à leur égard*. Paris: Mahaut.

1844b. *Second Mémoire: Situation comparée des colonies françaises et des colonies anglaises, par le baron Ch. Dupin*. Paris: Firmin Didot frères.

1844c. *Troisième Mémoire: Les colonies françaises devant la Chambre des Pairs par le baron Charles Dupin*. Paris: Firmin Didot frères.

1844d. *Constitution, histoire et avenir des Caisses d'épargne de France*. Paris: Firmin Didot frères.

1844e. *Discours au roi, prononcé le 1er janver 1844, en présentant à Sa Majesté les délégués et les membres des conseils coloniaux ou muncipaux des colonies, actuellement à Paris*. Paris; *Annales coloniales*, 1844, vol. 1, 145. Paris: Imprimerie royale.

1845a. *Discours de M. le Bon Charles Dupin, prononcé à l'assemblée de l'Association de fabricants et artisans pour l'adoption des orphelins des deux sexes*. General annual session, 6 July 1845. Paris: Firmin Didot frères, (n.d.)

1845b. *Statistique des Caisses d'épargne, crises éprouvées depuis l'origine jusqu'à la crise actuelle*. Paris: Panckouche.

1847a. *Du travail des enfants qu'emploient les ateliers, les usines et les manufactures, considéré dans les intérêts mutuels de la société, des familles, et de l'industrie*. (Parts 1 and 2). Paris: Bachelier. 1840–1847.

1847b. *Rapport pour les dépenses de l'Algérie*. Paris: Didot.

1847c. *Travaux et bienfaits de M. le baron Benjamin Delessert*. Paris: Firmin Didot frères.

1848. *Enseignement et sort des ouvriers de l'industrie avant, pendant et après 1848*. Lecture given at the conservatoire, 17 December. Paris: Pagnerre et Paulin.

1849a. *Situation actuelle de l'industrie française*. Paris.

1849b. *Eloge de Gaspard Monge*. Paris: Firmin Didot.

1850a. *Rapport sur le commerce de l'Algérie*. Paris: Firmin Didot.

1850b. *Great exhibition of the works of all nations, 1851. Suggestions addressed to the producers and manufacturers of France*. London: T. Bretelli.

1852. *Industries comparées de Paris et de Londres*. Paris: Firmin Didot frères.

1853a. *Société d'encouragement pour l'industrie nationale: Eloge de M. le comte Charles de Lasteyrie*. Paris: Veuve Bouchard-Huzard.

1853b. *Société d'encouragement pour l'industrie nationale: Funérailles de M. Pecquer, membre de la Société*. Paris: Veuve Bouchard-Huzard.

1856. *Société pour l'instruction élémentaire. Discours sur les progrès accomplis dans l'instruction élémentaire de la France*. Speech at a public session at the Sorbonne, 27 July. Paris: S. Raçon.

1857. *Canal maritime de Suez*. Paris: H. Plon.

1865. *The Baron Dupin's report to the emperor of the French upon the progress in the arts and sciences in Massachusetts and other states of New England* (with a preface and supplement by E. H. Derby). Boston: Wright and Potter.

1867a. Les forces productives des nations depuis 1800 jusqu'à 1851. In *Travaux de la commission française sur l'industrie des nations*, 12 vols., vol. 12. Paris: Commission française sur l'industrie des nations. 1854–1867.

1867b. *Force productive des nations*. Paris: Imprimerie impériale.

Endnote

1. George Canning (1776–1827), British statesman, briefly prime minister in 1827.

Secondary Bibliography

Ames, Glen J. 1996. *Colbert, mercantilism, and the French quest for Asian trade*. Dekalb: Northern Illinois University Press.

Anceau, Eric. 2009. Charles Dupin et la question sociale. Trois éclairages parlementaires (1840–48, 1848–49, 1864–68). In *Charles Dupin (1784–1873)*, edited by C. Christen and F. Vatin, 253–270. Rennes: Presse Universitaire de Rennes.

Bakewell, Sarah. 2000. The reanimators. In *Fortean Times*, edited by David Sutton, 139. London: Dennis Publishing.

Ballot, Charles. 1923. *L'introduction du machinisme dans l'industrie française*. Paris: Marquant, F., Rieder et Cie.

Barlow, Peter. 1817. *An essay on the strength and stress of timber*. London: J. Taylor.

Baron, Romain. 1978. La jeunesse et les années de formation de Dupin aîné (1783–1806). In *Mémoires de la Société académique du Nivernais*, 63–64. Nevers: Imprimerie de la Tribune.

Belhoste, Bruno. 2009. Charles Dupin et l'héritage de Monge en géométrie. In *Charles Dupin (1784–1873)*, edited by Carole Christen and François Vatin, 81–98. Rennes: Presse Universitaire de Rennes.

Benoît, Agnès. 2009. Le passeur des deux rives? La Grande-Bretagne dans l'action politique de Charles Dupin, 1814–1835. In *Charles Dupin (1784–1873)*, edited by Carole Christen and François Vatin, 53–66. Rennes: Presse Universitaire de Rennes.

Berland, Just. n.d. Histoire de l'Ecole de Châlons. Archives de la Marne.

Bertin, Louis François and Bertin, Pierre-Louis. 1789–1944. *Journal des débats politiques et littéraires*. Paris: Bertin.

Bertrand, Joseph. 1888. Eloge historique de Pierre-Charles-François Dupin. In *Mémoires de l'Académie des sciences*, vol. 44, xxiv. Paris: Gauthier Villars.

Billiet, M. 1825. *Du commerce, des douanes et du système des prohibitions; considéré dans ses rapports avec les intérêts respectifs des nations. Augmenté par M. Marie du Mesnil*. Paris: L. Bouchard.

Blaug, Mark. 1962. *Economic theory in retrospect*. Homewood, IL: R. D. Irwin. Revised 5th edition published 1997.

Blomfield, Charles J. 1814. A review in English of Dupin's Essais sur Démosthenes. *Museum criticum* 1, no. 4 (October): 563–566.

Blum, Jerome, Rondo Cameron, and Thomas G. Barnes. 1966. *The European world since 1815: Triumph and transition*. Boston: Little, Brown.

Bosset, Charles P. 1821. *Parga and the Ionian Islands*. London: John Warren.

Bourgoign, Victor A. de. 1971. Adieu à la Marine. *Les Annales du pays Nivernais* 1:3–11.

Bradley, Margaret. 1974. The Ecole Polytechnique, 1790 to 1830: Organisational changes and students. MPhil thesis, Leeds University.

———. 1975. Scientific education versus military training: The influence of Napoleon Bonaparte on the *Ecole polytechnique*. *Annals of Science* 32, no. 5: 415–449.

———. 1976a. An early science library and the provision of textbooks: The *Ecole Polytechnique*, 1794–1813. *Libri* 26, no. 3: 165–190.

———. 1976b. The facilities for practical instruction in science during the early years of the Ecole Polytechnique. *Annals of Science* 33, no. 5: 425–446.

———. 1976c. A scientific education for a new society: The *Ecole Polytechnique*, 1794–1830. *History of Education* vol. 5:11–24.

———. 1979. The financial basis of French scientific education and scientific institutions in Paris, 1790–1815. *Annals of Science* 36, no. 5: 451–491.

———. 1981. Franco-Russian engineering links: The careers of Lamé and Clapeyron, 1820–1830. *Annals of Science* 38, no. 3: 291–312.

———. 1992. Engineers as military spies? French engineers come to Britain, 1780–1790. *Annals of Science* 49:137–161.

———. 1994. Bonaparte's plans to invade England in 1801: The fortunes of Pierre Forfait. *Annals of Science* 51:453–475.

———. 1998. *A career biography of Gaspard Clair François Marie Riche de Prony, bridge-builder, educator and scientist*. New York: Mellen.

———. 2005. *Daniel Lescallier, 1743–1822: Man of the sea—and military spy*? New York: Mellen.

———. 2008. Dupin et la Grande Bretagne: La première visite. *Bulletin de la SABIX* 42 (April): 30–38.

Bradley, Margaret, and Fernand Perrin. 1991. Charles Dupin's study visits to the British Isles, 1816–1824. *Technology and Culture* 32, no. 1 (January): 47–68.

Brisson, Barnabie. 1818. *Notice historique sur Gaspard Monge*. Paris: Plancher.

Buret, Eugène. 1840. *De la misère officielle des classes laborieuses en France et en Angleterre*. Paris: Paulin.

Butenval, Charles A. 1876. *Tarif général des douanes, 1787–1791*. Paris: Guillaumin.

Chandru, Vijaya, Debasish Dutta, and Christoph M. Hoffmann. 1989. On the geometry of Dupin's cyclides. *Quarterly Journal of Pure and Applied Mathematics* 12:148–165.

Chappert, André. 1977. *E. L. Malus et la théorie corpisculaire de la lumière*. Paris: Vrin.

Charléty, Sébastien. 1921. *La Restauration*. Vol. 4, *Histoire de France Contemporaine*, edited by Ernest Lavisse. Paris: Hachette.

Chartier, Roger. 1992. La ligne Saint-Malo-Genève. In *Les lieux de mémoire: La France*, vol.1, Conflits et partages, edited by Pierre Nora, 738–775. Paris: Gallimard.

Chatelle, Albert. 1943. *Le Musée de la Marine*. Paris: Institut maritime et colonial.

Chatzis, Konstantinos. 2009. Charles Dupin, Jean-Victor Poncelet et leurs mécaniques pour 'artistes' et ouvriers. In *Charles Dupin (1784–1873)*, edited by Carole Christen and François Vatin, 99–114. Rennes: Press Universitaire de Rennes.

Chevalier, Michel. 1837. *Lettres sur l'Amérique du Nord*. Paris: Charles Gosselin.

Christen, Carole. 2009a. Charles Dupin, propagandiste des caisses d'épargne sous la monarchie de juillet. In *Charles Dupin (1784–1873)*, edited by Carole Christen and François Vatin, 235–252. Rennes: Press Universitaire de Rennes.

———. 2009b. Introduction. In *Charles Dupin (1784–1873)*, edited by Carole Christen and François Vatin, 11–28. Rennes: Press Universitaire de Rennes.

Christen, Carole, and François Vatin, eds. 2009. *Charles Dupin (1784–1873)*. Rennes: Presse Universitaire de Rennes.

Clapham, John H. 1921. *The economic development of France and Germany 1815–1914*. Cambridge: Cambridge University Press.

Clément, Jean-Pierre. 1846. *Vie de Colbert*. Paris: Imprimerie royale.

Clerk Maxwell, James. 1846. On the description of oval curves and those having a plurality of foci. *Proceedings of the Royal Society of Edinburgh* 2.

———. 1867. On the cyclide. *Quarterly Journal of Pure and Applied Mathematics*, 11:111–126.

Colladon, Jean-Daniel. 1893. *Souvenirs et mémoires*. Geneva: Droz.

Comberousse, Charles. 1879. *Histoire de l'Ecole centrale des arts et manufactures*. Paris: Gauthier-Villars.

Conservatoire national des arts et métiers, 1970. *Cent-cinquante ans de haut enseignement technique au Conservatoire national des arts et métiers*. Paris: Boutry Daumas.

Cooper, Caroline C. 1984. The Portsmouth system of manufacture. *Technology and Culture* 25 (April): 182–225.

Cordingly, David. 2003. *The Billy Ruffian: The Bellerophon and the downfall of Napoleon*. London: Bloomsbury.

Corro, Armando V., and Keti Tenenblat. 2004. Ribaucour transformations revisited. *Communications in Analysis and Geometry* 12, no. 5: 1055–1082.

Courrier, Paul-Louis. 1837. *Choix des historiens grecs avec notices biographiques par J. A. C. Buchon*. In *Panthéon littéraire*, vii–viii. Paris: Vve J. Renouard pour la Société de l'histoire de France.

Dajczer, Marcos, Luis A. Florit, and Ruy Tojeiro. 2007. The vectorial Ribacour transformation for submanifolds and applications. *Transactions of the American Mathematical Society* 359, no. 10: 4977–4997.

Delatte, Jean. 1982. Les Arts-et-Métiers: 200 ans d'histoire. *Arts et Métiers Magazine, Arts et Métiers Magazine* (May–June).

Démier, Francis. 2009. Charles Dupin: Un libéralisme sans doctrine? In *Charles Dupin (1784–1873)*, edited by Carole Christen and François Vatin, 165–176. Rennes: Press Universitaire de Rennes, 165–176.

DiLorenzo, Thomas, J. 1999. Frédéric Bastiat (1801–1850): Between the French and the marginalist revolutions. In *15 great Austrian Economists*, edited by Randall G. Holcombe, 59–70. Auburn, AL: Ludwig von Mises Institute.

Dulaure, Jacques-Antoine. 1821. *Histoire physique, civile et morale de Paris*. Paris: Furne et cie.

Duméril, André-Marie C. (reporter) 1827. Compte rendu sur les ouvrages de Charles Dupin. *Mémoires de l'institut national des sciences et des arts* 70, no. 28 (August).

Dunham, Arthur L. 1955. *The Industrial Revolution in France, 1815–1848*. New York: Columbia University Press

Dupin, André-Marie J. J. 1826. *Du droit d'aînesse*. Paris: Paul Ledoux.

———. 1853. *Défense de l'évêché de Nevers Le Morvan*. Paris: Plon-Frères.

Dupin, Claude. 1804. *Mémoire statistique du Département des Deux-Sèvres*. Paris: Imprimerie de la République.

Dupont, Louis. 1786. Mémoires et documents, Angleterre. Manuscript 65, fol. 88, third mémoire of Dupont, Archives Etrangères (Archives of the Ministère des Affaires Etrangères), Paris.

Duvoir, Léon. 1901. *Recherches des tendances interventionnistes chez quelques économistes libéraux français de 1830 à 1850*. Paris: Rousseau.

Dyer, Gwynne. 2007. Population bomb still ticking away. *NZ Herald*, 20 March 2007.

Ecole polytechnique. 1895. *Livre du centenaire (1794–189)*. Paris: Gauthier-Villars.

Estaunié, Edouard, ed. 1932. *L'Ecole polytechnique à travers l'histoire*. Paris: Gauthier-Villars.

Falguerolles, Antoine de. 2009. Charles Dupin, statisticien. In *Charles Dupin (1784–1873)*, edited by Carole Christen and François Vatin, 115–128. Rennes: Press Universitaire de Rennes.

Farrère, Claude. 1934. *Histoire de la Marine française*. Paris: Flammarion.

Faucher, Léon. 1845. *Etudes sur l'Angleterre*. Paris: Guillaumin.

Ferreiro, Larrie D. 2007. *Ships and science: The birth of naval architecture in the scientific revolution, 1600–1800*. Cambridge, MA: MIT Press.

Flinn, Michael W. 1926–1927. The travel diaries of Swedish engineers of the eighteenth century as sources of technological history. *Transactions of the Newcomen Society* 7:95–109.

Fohlen, Claude. 1967. *Le travail au XIXe siècle*. Paris: Presses Universitaires de France.

Fourcy, Ambroise. 1828. *Histoire de l'Ecole polytechnique*. Paris: Chez l'auteur.

Fox, Robert. 1974. Education for a new age: The *Conservatoire des Arts et Métiers*, 1815–1830. In *Artisan to Graduate*, edited by D. Cardwell, 23–38. Manchester: Manchester University Press.

———, ed. 1978. *S. Carnot, Réflexions sur la puissance motrice du feu*. Paris: Vrin.

———. 1985. From Corfu to Caledonia: The early travels of Charles Dupin, 1808–1820. In *The Light of Nature*, edited by J. D. North and J. J. Roche, 303–320. Dordrecht: Kluwer.

———. 2009. Le 'vrai' Charles Dupin. In *Charles Dupin (1784–1873)*, edited by Carole Christen and François Vatin, 7–11. Rennes: Press Universitaire de Rennes.

French Ministry of Education. 1970. *1820–1970, cent cinquante ans de haute enseignement technique au Conservatoire des arts et métiers*. Paris: Ministère de l'éducation nationale.

Gallois, Louis G. G. de. 1818. Des chemins de fer en Angleterre. *Annales des mines* 3:137–144.

Geikie, Archibald, Sir. 1907. (edited and translated with notes, the work of Barthélémie Faujas de Saint-Fond), *A Journey through England to the Hebrides in 1784*. 2 vols. Glasgow: High Hopkins. Originally published as Faujas de Saint-Fond, Barthélémie. 1797. *Voyage en Angleterre, en Ecosse et aux îles Hébrides*. 2 vols: Paris: Jansen.

Gilbert, Keith 1965. *The Portsmouth Block-Making Machinery*. London: H. M. Stationery Office.

Gillispie, Charles C. 1997. *Pierre-Simon Laplace 1749–1827*. Princeton, NJ: Princeton University Press.

Grattan-Guinness, Ivor. 1990. *Convolutions in French mathematics, 1800–1840*. Basel: Birkhauser.

———. 1998. *The Norton history of the mathematical sciences*. New York: Norton.

———. 2008. Les discours d'ouverture de Charles Dupin au Conservatoire. *SABIX* no. 42 (April): 26–30.

Graves, Donald E. 2006. The mysteries of the school at Woolwich: Royal artillery training and texts, c. 1750–1850. Gareth Newfield, ed., *The War of 1812 Magazine* 2 (February). http://serials.abc-clio.com.

Grelon, André. 1996. La naissance de l'enseignement supérieur industriel en France. In *Quaderns d'historia de l'Enginyeria*, 40–60. Barcelona: University of Barcelona.

Gueslin, André. 2009. Conclusion générale. In *Charles Dupin (1784–1873)*, edited by Carole Christen and François Vatin, 271–278. Rennes: Presse Universitaire de Rennes.

Guizot, François P. G. 1864. *Mémoires pour servir à l'histoire de mon temps*. Paris: M. Lévy frères.

Guyon, Louis. 1818. *Eloge funèbre de M. Monge*. Paris: Plancher.

Hachette, J. N. 1804–1816. *Correspondance sur l'Ecole impériale polytechnique*. Paris: Klostermann.

Hampshire, Denise. 2006. *Living and working in France*. London: Survival Books.

Harris, John R. 1985. Industrial espionage in the eighteenth century. *Industrial Archaeology Review* 7, no. 2.

———. 1998. *Industrial espionage and technology transfer*. Aldershot: Ashgate.

Harrison, Ross. 1995. Jeremy Bentham. In *The Oxford Companion to Philosophy*, edited by Ted Honderich, 85–88. Oxford: Oxford University Press.

Henderson, William O. 1966. *Industrial Revolution on the Continent, 1800–1914*. London: Routledge.

———. 1982. Friedrich List, and the French protectionists. *Journal of Institutional and Theoretical Economics* (Hallman, Berlin): 262–275.

Hollander, Samuel. 1979. *The economics of David Ricardo*. Toronto: University of Toronto Press.

———. 1997. *The economics of Thomas Robert Malthus*. Toronto: University of Toronto Press.

———. 2001. *The origins of David Hume's economics*. London: Routledge.

Jarrige, François. 2009. 'Nous ne sommes ni vos élèves ni vos amis': Charles Dupin face à la presse ouvrière sous la monarchie de juillet. In *Charles Dupin (1784–1873)*, edited by Carole Christen and François Vatin, 223–234. Rennes: Presse Universitaire de Rennes.

Jars, Gabriel. 1774–1781. *Voyages Métallurgiques*. 3 vols. Lyon: Regnault.

Jeremy, David. 1981. *Transatlantic Industrial Revolution*. Cambridge, MA: MIT Press.

Junges, Catherine. 2009. Charles Dupin, ingénieur du Génie Maritime. In *Charles Dupin (1784–1873)*, edited by Carole Christen and François Vatin, 43–52. Rennes: Press Universitaire de Rennes.

Karila-Cohen, Pierre. 2009. L'impossible ligne Saint-Malo-Genève, des opinions? Un essai de 'statistique politique' sous la Restauration. In *Charles Dupin (1784–1873)*, edited by Carole Christen and François Vatin, 129–148. Rennes: Press Universitaire de Rennes.

Lacaine, Victor, and Charles Laurent, eds. 1847. *Biographies et nécrologies des hommes marquants du 19e Siècle*. 4 vols. Paris: A la Direction, rue St Dominique, St Germain, No. 55.

Lambert, Andrew. 1991. *The last sailing battlefleet: Maintaining naval mastery, 1815–1850*. London: Conway.

———. 2006. Sir Robert Seppings and the British response to the lessons of Trafalgar. *Technology of the Ships of Trafalgar*, edited by Francisco Fernàndez-Gonzàlez, Larrie D. Ferreiro, and Horst Nowaski, Section 18:1, 18–27. Madrid: Escuela Técnica Superior de Ingenieros Navales, Universidad Politécnica de Madrid.

Lancret, M. A. 1806. Développoïdes des courbes à double courbure. Paris: Académie des sciences.

Larousse, Pierre. 1866–1877. *Grand Dictionnaire Universel du XIXe Siècle*. Paris: Administration du Grand Dictionnaire

Lavisse, Ernest. 1921. *Histoire de la France contemporaine*. 10 vols. Paris: Hachette.

Leblanc, César N. L. 1830. *Choix de modèles appliqués à l'enseignement du dessin de machines avec un texte descriptif*. Paris: Conservatoire national des arts et métiers.

Le Mallier, Denise. 1971. *Le Roman des Dufaud*. La Charité sur Loire: Delayance.

Lemesle, Lucette le Van. 2001. Les économistes français et l'usage des modèles étrangers. *Revue d'histoire du XIXe siècle*, 1–13. http://rh19.revues.org.

Léon, Antoine. 1968. *Histoire de l'éducation technique*. Paris: Presses Universitaires de France.

Le Play, Frédéric. 1864. *La réforme sociale*. Paris: Dentu.

Lévêque, Charles. 1873. Discours aux funérailles de M. Dupin. *Séances et travaux de l'ASMP* 99:609.

List, Friedrich. 1841. *Das nationale system der politischen oekonomie*. Stuttgart: Cotta. Translated by Henri Richelot in 1857 as *Le système naturel d'économie politique*. Paris: Corbeil.

Littré, E. 1863–1873. *Dictionnaire de la langue française*. Paris: Hachette.

Lutun, Bernard, ed. 1994. *Mémoires du Baron Tupinier, directeur des ports et arsenaux: 1770–1851*. Paris: Desjonquières.

Malet, Albert, and Jules Isaac. 1922. *Histoire de France, 1741–1851*. Paris: Hachette.

Mannheim, Amédée. 1894. *Principes et développements de géométrie cinématique*. Paris: Gauthier-Villars et fils.

Mascart, Jean. 2000. *La vie et les travaux du Chevalier Jean-Charles de Borda (1733–1799)*. Paris: Sorbonne.

McWatters, Keith G., ed. 1993. *Chroniques pour l'Angleterre: Contributions à la presse britannique, 1827–1829, par Stendhal*. Grenoble: Presse Universitaire de Grenoble.

Mertens, Joost. 2004. The *Annales de l'industrie* (1820–1827): A technological laboratory of the industrial modernization of France. *History and Technology* 20 (June): 135–163.

———. 2010. The mere handicrafts: Ure's (1839–1853) dictionary compared with the Dictionnaire Technologique (1822–1835). In *Les techniques et la technologie entre la France et la Grande-Bretagne XVIIe-XIXe siècles*, edited by P. Bret, I. Gouzevitch, and L. Pérez, 277–286.

Missiessy, Edouard J. B. de. 1798. *Traité de l'installation des vaisseaux*. Paris: Imprimerie nationale.

Monge, Gaspard. 1799. *Géométrie descriptive*. Paris: Baudouin.

Montagne, Pierre. 1979. *Les cinq frères Deluret, Officiers de Guerre de la Révolution et de l'Empire*. Paris: Orfeuil.

Monzie, Anatole de. 1848. *Le Conservatoire du peuple*. Paris: Lefebvre, Lucien.

Moraud, Marcel. 1954. *Une Irlandaise libérale en France sous la Restauration, Lady Morgan 1775–1859*. Paris: Marcel Didier.

Morin, Arthur. 1873. Discours prononcé aux funérailles de M. le Baron Charles Dupin. Paris: Institut de France.

Morin, Marie-Renée. 1995. *Correspondance de Lamartine avec Charles Dupin et documents épistolaires*. Paris: Librairie Nizet.

Mouchelet, Emile, 1913. *Notice historique sur l'Ecole centrale des arts et manufactures*. Paris: H. Dunod et E. Pinet.

Mullié, Charles. 1852. *Biographie des célébrités militaires des armées de terre et de mer de 1789 à 1850*. Paris: de Surcy.

Oliver, Bette W. 2005. From libraries and culture to the Bibliothèque Nationale. *Libraries and Culture* 40, no. 3:455–459.

Palsky, Gilles. 1990. La naissance et la diffusion d'une méthode de cartographie quantitative: La carte teintée du baron Charles Dupin. *Bulletin du Comité français de cartographie* 125 (September): 5–11.

———. 1996. *Des chiffres et des cartes. La cartographie quantitative au X1Xe siècle*. Paris: Comité des travaux historiques et scientifiques.

———. 2007. Connections and exchange in European cartography: The example of 19th century statistical maps. Invited paper for the Sixth International Conference of the Brussels International Map Collectors' Circle, 16 November 2007.

Pares, Bernard. 1962. *A history of Russia*. London: Methuen.

Patterson, Elizabeth C. 1983. *Mary Somerville and the cultivation of science, 1815–1840*. The Hague: Martinus Nijhoff.

Payen, Jacques. 1998. The role of the Conservatoire des Arts et Métiers in the development of technology up to the middle of the nineteenth century. *History and Technology* 5, nos. 2–4:95–138.

Payne, David. 2008. The French army in the Great War. The Western Front Association. 10 October. http://www.westernfrontassociation.com.

Pernot, Hubert, ed. 1918. *Nos anciens à Corfou, souvenirs de l'aide-major Lamare-Picquot (1807–1814)*. Paris: Félix-Algan.

Pernoud, Régine. 1960. *Histoire de la bourgeoisie en France*. Paris: Sirey.

Perrin, Fernand. 1983. La vie et l'oeuvre de Charles Dupin (1784–1873), mathématicien, ingénieur et éducateur. Doctorate thesis, Ecole des hautes études en sciences naturelles.

Perrot, Michelle. 1976. Premières mesures des faits sociaux: les débuts de la statistique criminelle en France (1780–1830). In *Pour une histoire de la statistique*, 125–177. Paris: Institut national de la statistique et des échanges économiques. http://genes.bibli.fr/opac/index.php ?lvl=author-see&id=18148.

Phili, Christina. 2006. Jean Carandinos (1784–1834): Initiateur des mathématiques françaises en Grèce. *Archives internationales d'histoire des sciences* 56, no. 156–157 (June–December): 82–125.

Philips, Chris. 2000. Robert Woodhouse and the evolution of Cambridge mathematics. *History of Science* 44:69–91.

Ponteil, Félix. 1966. *Institutions de la France 1814–1870*. Paris: PUF.

Pothier, Francis. 1887. Histoire de l'Ecole centrale des arts et manufacture. Paris: Delamotte.

Ragey, Louis. 1970. In *1820–1970, 150 ans de haut enseignement technique au Conservatoire national des Arts-et-Métiers*, edited by George-Albert Boutry and Maurice Daumas. Paris: Ministère de l'Education nationale.

Robert, A., E. Bourloton, and G. Cougny. 1891. *Dictionnaire des parlementaires français*. Paris.

Robin, Charles. 1848. *Histoire de la révolution*. 2 vols. Paris: Eugène et Victor Penaud frères.

Robinson, Eric. 1958. International exchange of men and machines. *Business History* 1:3–15.

———. 1975. The transference of British technology to Russia. *Great Britain and Her World, 1750–1940*, edited by Barrie M. Ratcliffe. Manchester: Manchester University Press.

Rodger, Nicholas A. M. 2004. *The command of the ocean: A naval history of Britain, 1649–1815*. London: Allen Lane.

Rosen, Frederick. 2003. *Classical utilitarianism from Hume to Mill: Routledge studies in ethics and moral theory*. London: Routledge.

Ruffieux, Christiane. 2005. La naissance du concept de structure algébrique en Grande Bretagne dans la première moitié du 19ème siècle. PhD thesis, University of Geneva.

Saint-Fond, Barthélémie Faujas de. 1907. *A Journey through England to the Hebrides in 1784*. 2 vols. Edited and translated with notes by Sir Archibald Geikie. Glasgow: High Hopkins. Originally published 1797. See Geikie 1907.

Say, Jean-Baptiste. 1803. *Traité d'économie politique*. Paris: Horace Say.

———. 1815. *De l'Angleterre et des Anglais*. Paris: Bertrand.

Scull, Thomas. 2007. Tribute to QE2. *Ships Monthly*, December, 18–22.

Seppings, Robert. 1814. Mr Seppings on a new principle of constructing his majesty's ships of war. *Philosophical Transactions of the Royal Society*, 19 March, 285–302.

———. 1818. On the great strength given to ships of war by the application of diagonal braces. (Read, Royal Society, 17 November 1817). *Philosophical Transactions of the Royal Society*, 1 January 1818: 1–8.

Sherman, Dennis. 1974. The meaning of economic liberalism in mid-nineteenth-century France. *History of Political Economy* 6, no. 2:171–199.

Simond, Charles. 1900. *La vie parisienne à travers le XIX siècle: Paris de 1800 à 1900*. Paris: E. Plon.

Smith, Crosbie. 1976. Mechanical philosophy and the emergence of physics in Britain: 1800–1850. *Annals of Science* 33, no. 1 (January): 3–29.

Steiner, Philippe. 2009. Charles Dupin et Jean-Baptiste Say: Arithmétique politique et économie politique. In *Charles Dupin (1784–1873)*, edited by Carole Christen and François Vatin, 149–164. Rennes: Press Universitaire de Rennes.

Szostak, Rick. 1991. *The role of transportation in the Industrial Revolution: A comparison of England and France*. Montreal: McGill-Queen's University Press.

Tannery, Paul. 1887–1902. Charles Dupin. In *La grande encyclopédie*, edited by Gaston Gallimard, 31 vols., vol. 15:81. Paris: Gallimard.

Taton, René. 1951. *L'oeuvre scientifique de Monge*. Paris: Presses Universitaires de France.

Thompson, Edward P. 1963. *The English Working Class*. London: Penguin.

Thuillier, Georges. 1976. *Le corps préfectoral en Nivernais de 1814 à 1830.* Evreux: Floréal.

Timon [L. M. de la H. Cormenin, pseud.]. 1837. *Etudes sur les orateurs parlementaires*. Paris: Pagnèrre.

Todd, David. 2008. *L'identité économique de la France*. Paris: Grasset.

———. 2009. La liberté, la nation et les colonies dans la pensée économique de Charles Dupin. In *Charles Dupin (1784–1873)*, edited by Carole Christen and François Vatin, 177–190. Rennes: Presse Universitaire de Rennes.

Tousard, Louis. 1813. *American artillerist's companion.* 2 vols. Philadelphia: Bradford and Innskeep.

Tresca, Henri. 1873. Eloge du Baron Charles Dupin, membre de l'Académie des sciences, secrétaire général de la *Société d'Encouragement. Annales du Conservatoire des arts et métiers* 1, no. 10:244–259.

Tulard, Jean. 1970. *Nouvelle Histoire de Paris sous le Consulat et l'Empire*. Paris: Hachette.

———. 1982. *Le grand empire*. Paris: Albin Michel.

Turgot. 1769–1770. Réflexions sur la formation et la distribution des richesses. In *Ephémérides du citoyen*, edited by Abbé Nicolas Beaudeau, 12–56. Paris: Lacombe.

Vatin, François. 2008. *Morale industrielle et calcul économique dans le premier XIXe siècle: L'économie industrielle de Claude-Lucien Bergery*.Paris: L'Harmattan.

Verger, Jacques, ed. 1986. *Histoire des universités en France*. Toulouse: Privat.

Villiers, Patrick. 2005. Et si Trafalgar n'avait jamais existé, ou le mythe de la victoire decisive. *Revue maritime* 472 (May): 1–8.

Werner, Alex. 2010. Scaling London's early 19th century docks, bridges and manufactories: Charles Dupin's writings and technological exchange. In *Les échanges techniques entre la France et l'Angleterre (XVIe-XIXe siècles)*, edited by P. Bret, I. Gouzevitch and L. Perez, 199–208. Paris: CNRS.

Wolkowitsch, Mauricz. 1960. *L'Economie des transports dans le centre et le centre ouest de la France*. Paris: Sedes.

Woodhouse, Robert. 1810. *A treatise on isoperimetrical problems and the calculus of variations*. Cambridge: J. Smith.

Young, Arthur. 1792. *Travels in France during the years 1787, 1788 and 1789*. London: Betham Edwards.

Index

www.ingramcontent.com/pod-product-compliance
Lightning Source LLC
LaVergne TN
LVHW092116300926
849161LV00002B/151
* 9 7 8 1 6 0 4 9 7 7 5 1 6 *